Even More Schedule for Sale

Advanced Work Packaging, for Construction Projects

GEOFF RYAN P. M. P.

AuthorHouse™
1663 Liberty Drive
Bloomington, IN 47403
www.authorhouse.com
Phone: 1 (800) 839-8640

Published by AuthorHouse 11/13/2017

ISBN: 978-1-5462-0430-5 (sc)
ISBN: 978-1-5462-1701-5 (hc)
ISBN: 978-1-5462-0431-2 (e)

Library of Congress Control Number: 2017912627

Print information available on the last page.

This book is printed on acid-free paper.

authorHOUSE®

As you may know this is the second book that I have developed on the subject of Project Management, the first: 'Schedule For Sale, Workface Planning for Construction Projects' explored the subject of Workface Planning (WFP) as a means for the industry to work package our way to managing mega projects. It was the equivalent of baby steps. A learning experience where we tried new things and then stood back to see the effect.

I understand that learning to walk is one of the most effective learning experiences that we ever have, because the cycle between cause and effect is immediate. Our learning cycle with WFP was not quite as quick, but we certainly did get some blood noses from it.

And now here we are as an industry 10 years later and we figure that we have nailed this WFP thing (walking) and its time to learn to run, welcome to Advanced Work Packaging and Information Management. Truth is that we are still a little shaky on WFP, but mankind is not the sort to let fear or common sense hold us back, so here we go.

Right from the start of the development of the WFP model I remember the other members of the Construction Owners Association of Alberta (COAA) WFP committee talking about the need for support from Engineering and Procurement. At the time, I think that we agreed that we would have to let it slide while we leveled our focus on getting WFP off the ground. Well it didn't take long for the first couple of projects in 2006 to identify that the need for engineering and procurement support was real and quite obviously the next target for development.

The team at COAA spent the next couple of years helping the industry get their heads around what WFP was and luckily for us (and the industry) it caught the eye of the Construction Industry Institute (CII). The CII have a very effective model for process exploration and development that uses University Professors and Students to create studies that address specific industry problems. Recognizing that there was an opportunity to explore and develop processes that would enhance the effectiveness of WFP, CII established Research Team 272 with the mandate to flush out the best practices around Engineering and Procurement that would support the execution of WFP during construction.... and Advanced Work Packaging (AWP) was born.

Several years later, the industry had applied the CII model for AWP on several projects and incorporated the preparation of Engineering and Procurement deliverables with the COAA model for WFP. A second CII committee then set out to prove the results of the process and the results of those years of research was the catalyst that drove the process of AWP to be identified as an industry Best Practice.

Not quite Olympic pace marathons yet, but the early adopters certainly have figured out how to run sprints.

For those of you that haven't got a clue what I just said, (the Folks that didn't read the original Schedule For Sale), don't worry, we will get into the details of each of these subjects later in the book. Right now, I just wanted you to know that there have been a lot of smart people do a lot of hard work, risking their reputations and careers to get this process to where it is today. I have the opportunity and the privilege to bring their hard work together in this book so that the rest of the industry can benefit from their vision and determination.

Advanced Work Packaging, Information Management and Workface Planning for Construction Projects

Contents

Introduction: 1
Acknowledgements: 3
Chapter 1: Why Change? 4
Chapter 2: What is AWP, IM & WFP 9
Chapter 3: Return on Investment & Benefits 12
Chapter 4: AWP Quick Start Guide 17
Chapter 5: Advanced Work Packaging 28
Chapter 6: Information Management 54
Chapter 7: Workface Planning 73
Chapter 8: Productivity Management 88
Chapter 9: AWP Testimonies 110
Chapter 10: The Future: 118
Summary and Links: 125

Introduction:

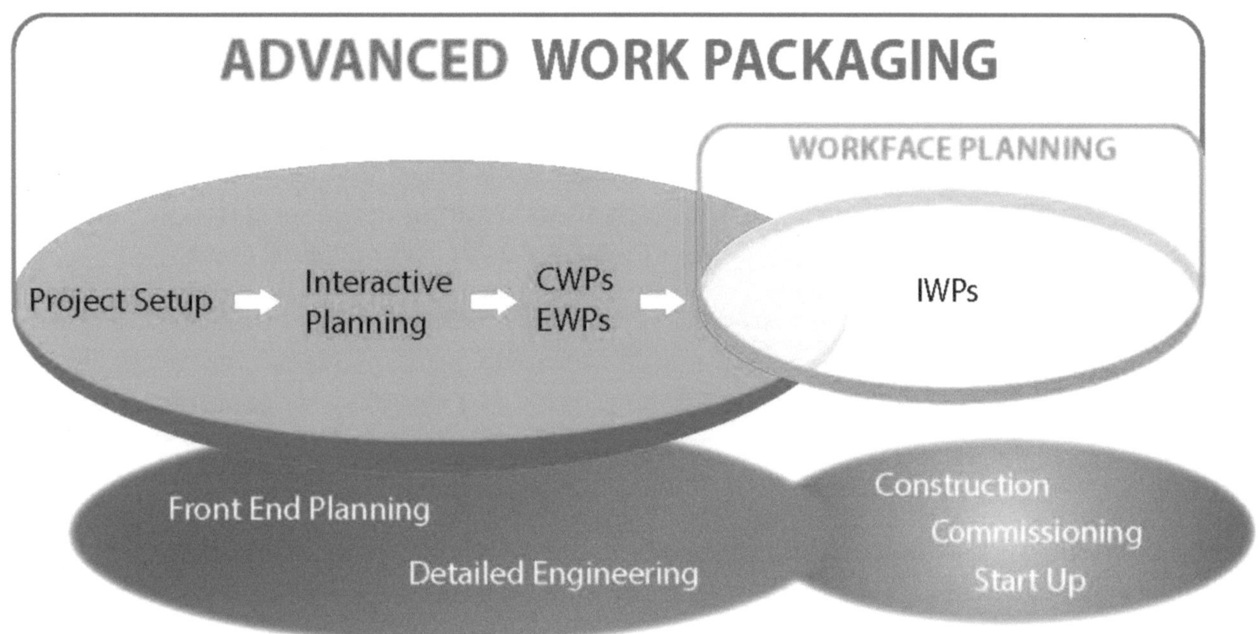

Reprinted with permission from CII

Let's start with the CII surfboards, I find that occasionally academics are very good at explaining complex issues in very simple terms, this is certainly the case with these surfboards.

The process is shown here as two overlapping processes that show preparation and execution based on the development of plans (EWPs, CWPs and IWPs)

While the entire process is referenced as Advanced Work Packaging (an enhanced level of project planning) the front-end preparation (the blue surfboard) is also commonly referenced specifically as Advanced Work Packaging (it happens first). For the purpose of this book AWP covers the activities on the blue surfboard.

There is also another element that we have identified as being distinctive and worthy of its own surfboard: Information Management. The process is both unique and interwoven with AWP and WFP and can be described as the glue that holds them together:

Information Management

The guiding principle of AWP, IM and WFP is that we can significantly reduce our construction schedules, optimize our safety and quality performance, minimize cost and make construction projects predictable by being organized. That organisation comes in the form of constraint free Installation Work Packages that are one week of work for each Foreman and crew.

The idea of being ready before you start work is not really rocket science, but it is actually very, very difficult to do. So, let me share a lesson taught to me by a Saskatchewan farmer: 'Pull hard, comes easy'. Don't set off on this journey thinking that it is a doddle. The best results that we have seen in the industry have come from teams that have survived a project train wreck. They know that they have to pull hard.

Having said that, projects are doing this and achieving outstanding results. This means that today certain pockets of the industry have a predictable probability of project success, driven by a combination of experience, process, preparation, effort and perseverance. Over the next couple of years we will get to the point where the rest of the industry has a common understanding of how to apply AWP, IM and WFP. Then we will be able to see failed projects for what they truly are: a lack of design and effort (not knowledge).

The purpose of the book is to help create that common platform of understanding by giving everybody in the industry access to the best-known methods and processes, as we did in the original version.

Your obligation now that you have picked up this book, is to learn as much as you can from our experiences and use them to create your own model for change and optimal project results.

This book has lots more ideas and processes that complement the ideas in the first version. This substantiates the concept that as an industry, we can learn from each other and resolve new issues as they appear. We can expect that to happen again, if you take it upon yourself to become a change agent and push that envelope.

Most of the processes addressed in the book overlap into several areas of influence, the squad check used to turnover the CWP from the Construction Management team to the Construction Supervisor is a good example. It is mentioned in three different places in the book where it's description supports the rest of the framework in the chapter. As such you will probably catch yourself reading passages that are similar to ones that you have already covered. While the primary purpose is to show context with the rest of the passage it is also a sure sign that it is important and worth noting.

Acknowledgements:

The process that I used to develop the book was that I spent the last 20 years identifying a list of Doers and then characterized them by WFP, IM and AWP applicators. From that batch, I selected the ones that had real battle scars and then I asked this surprisingly big group of industry experts to help me develop the book. I did this by sending them a chapter a month and asking them to add their real-world experience and vision to the pages. My own understanding of WFP, IM and AWP has been significantly enhanced by these folks and they have given the book a life of its own. I truly believe that the following pages represent the best-known methods and experience that the industry has to offer.

I also offered the same group of people the opportunity to develop some testimonials that you will find at the back of the book. Their words are heart felt and offer an insight into the passion and dedication that these subject matter experts have for process improvement and the continuous evolution of our industry.

Chapter 1: Why Change?

Take a few moments and imagine the world of construction in the future. First think of what it will be like in 10 years, then 20 years......... and now 50 years..........

Changes driven by technology are probably a big part of what you imagined. If you stop to think about the changes that we have seen in our life time and then think about how it has impacted your world and habits, you will see that we truly are living in a paradigm shift.

Just think about the way that you get your news now. The newspaper in the morning and the television in the evening were the sole source of information for me as a young man. Now my phone pings with real time text and video based on which global news stream I have subscribed to. I haven't picked up a newspaper in many years and I think that my television only broadcasts sports now. That is a paradigm shift in our ability to process data when you consider my grandfather who looked forward to reading a weekly newspaper and going to the cinema once a month.

And the pace of acceleration will only get faster

Moore's Law projected the speed of change in 1965: "*The number of transistors in a dense integrated circuit will double every two years*". It was later revised to 18 months and has proven to be true over the last 50 years.

Let me have a stab at what construction projects might look like in 20 years: (2037, I'll be 74 and hopefully slowing down a golf game somewhere) "Predictable" will be a word that is only used by the newbies, like "out of the box" is used now. We have long since passed the idea that we don't know what projects will cost or how long they will take, or if we should build them at all.

Picture this:

A market trend shows that plastic is trending up and the world supply will not be able to meet demand in the next 10 years, so a chemical company executive pulls up his "build it" app and plugs in 100,000-ton plant in southern India (because it has the best shipping routes to markets) and the app checks the work load of steel mills, plastic production, fabrication shops and module yards around the world and gets an average price and schedule for steel and pipe. Then using a standard design 3D model for a 100,000-ton production plant it makes a list of the equipment, valves, instruments and special equipment that will be required and checks the supply chain for price and schedule.

Then milliseconds later the price and schedule pop up on the executive's watch and a green light indicates that based upon todays markets for the product and the supply chain of materials and labor, the project has a good ROI and a 78% probability of success.

Unfortunately for the executive his company is trying to establish itself as a blue-chip investment and they require an 80% probability of success to be funded.

At the same time, a kid living in his mother's basement in Kazakhstan has started an upstart chemical company, Chemtronics, named after his favorite transformer, funded by tens of Millions of $10 investors. He sees the same opportunity and gives it the green light and a project is born. He posts the project on the construction version of "go fund me". A realtor responds with a plot of land that has dock and high way access along with government sponsored tax incentive plans. An Engineering company offers an off the shelf design for a 100,000-ton plant. The local unions show that they have the available workforce and an AWP certified project management company respond with a fixed price guarantee (with some incentives for early delivery).

If you think that is a little far-fetched imagine a free app that can tell you the exact location, altitude, speed and landing schedule for all 10,000 aircraft that are in the sky right now, (there are lots of them).

The key is data. One of the reasons that our current model for industrial construction is not predictable is because our methods and data are not standardized.

Needs drive change. The industrial construction industry is driven by the need for a Return On Investment (ROI). The more we know and more finite our investment strategies become the more we will demand predictable outcomes. Plants are built or not based upon your own investment strategies and your need for a ROI. The world needs predictable ROIs to support the demands of the investment markets. Right now, an investment in a construction project is a wild ass gamble and that needs to change, because we have lots of stuff that must get built and we need to be able to make decisions based on a probability index, not a sales pitch.

Coming back to earth and the problem at hand:

In the same 50 years where we doubled the capacity of data management every 18 months, steel construction transitioned from using rivets to bolts.

If we put our heart into it I'm sure that we could design steel connections that don't need to be bolted or welded. Have a look at the steel connections used for earthquake proof buildings.

Which proves to me that the changes that we will see in construction are influenced much less by what we could do (technology) and much more by what we must do (process). These changes in contractor processes are typically influenced by desperation, survival, regulations or accountability (stick motivators) and sometimes by profit (carrot motivator). Ultimately the Owner decides which of these influences becomes the driver for change. The issue thus far is that the owners don't always know what changes they want or what is on the table and even if they do, they don't understand how to formulate the outcome.

Most owners will have a broad desire for 'on-time & on-budget' but what does that mean? Who's time and who's budget? What they really want is predictability and the confidence to know that the cost and schedule were appropriate. The industry has been trying for years to secure this outcome with contracts that are tailored to drive these behaviors, without much success.

Without getting too far off track, let's have a quick look at three primary contracting strategies, just to set some parameters:

Lump-sum and unit-rate contracts: We get the floor mixed up with the roof, we think that hard priced contracts represent the most that we will pay for a product (the roof). It's not, it's the floor. The price on the bottom of that contract is the least that you will pay, the price will definitely go up from there. And the contractor spends lots of time and effort focused on how to make that happen, rather than building your project. Importantly these contracts do not shift the risk from the owner to the contractor, the risk just gets added to the price.

Time and materials (milk and honey) contracts: A necessary evil in a fast track delivery world of schedule driven projects where 100% engineering and materials costs you too much schedule. The math works if you manage it and it gives us the flexibility to design on the fly, but the management team needs to be focused on facilitating the project, not cost reduction through siloed budgets (choke management).

Triple P: Public/Private Partnerships are an example of contracts that do appear to drive the right behaviors. This contract establishes the contractor as the owner, who finances and manages the design and build of a facility (Hospitals and highways), renting it back to the owner for a fixed 25-50-year term. The motivators for schedule, cost and quality are all in the right place, but you need a contractor that understands how to be productive, who also has deep pockets.

One thing that I know for sure (thanks to Stephen Covey) is that your overall strategy with any contract type must be based on Win-Win. In my 25 years, I have never seen Win-Lose, just Win-Win or Lose- Lose.

The outcome from any good contract is that the contractor makes a reasonable profit and that the owner gets a functioning product that is fit for purpose. Usually that means that it makes money (schedule driven) or provides a service (cost driven).

Coming back to our desire for change, the message is that in lump-sum, unit-rate or time & materials contracts the Owner must be the change agent. They are the stakeholder that stands to gain most from productive execution and the synchronization between Engineering, Procurement and Construction. In order to do this, they must identify their expectations and create a project management environment that governs the articulation of change.

In short: Owners must be in the business of management, not just spectators at a football game screaming criticisms.

Where are we today? and Where do we want to be?

Today we are in a world where only a small percentage of mega projects reach their business objectives and hardly any land within 10% of their cost and schedule targets.

30 years of global tool time studies have shown us that the average worker spends less than 4 out of every 10 hours doing what we want them to do.

This means that we know that our productivity is poor and that our best estimates for cost and schedule are unreliable.

What we want, is to be able to say it and do it, develop real estimates and then execute projects predictably.

Therefore, the answer to 'Why should we change?" is because we can, and if we don't then one of those forces that we talked about earlier will appear: Desperation, survival, regulations or accountability and maybe even the need for profit.

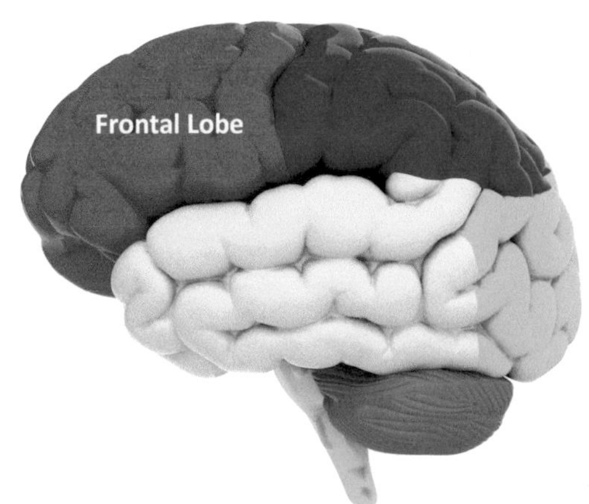

The frontal lobe is that portion of the brain beneath your forehead that is responsible for the management of dopamine amongst other things. The distribution of dopamine influences short term memory, motivation and planning. And it is this ability to visualize a future state that separates us from most of the rest of the animal kingdom. What does that mean? It means that the reason that we run the planet is because we have the ability to plan and then to take actions based upon the probability of future outcomes. (that and we have thumbs)

That means that with the gentle stimulation that you just experienced combined with your hard-wired ability to envision outcomes, you can create a pretty good picture of where our industry is going.

We probably will see advanced technologies and systems introduced that completely change the way that we approach projects. Our expectations for results and tolerance for risk will be very tight and we will be aghast when we look back at project results from the turn of this century. Much like we do now when we look at the average death rates from big construction projects in the past.

The thing to remember is that you have a role in this process, we are all players on the main stage, with the course of history and the pace of change in our hands. The same instinctual drivers that led us out of the trees and onto the surface of Mars are still imbedded in each of us. We are hard wired to continuously improve.

On top of that I also have a theory about the frontier people who populate the ranks of our construction projects, that goes something like this:

Some time ago you were at home in your place of birth and woke up one morning, looked around and said: 'There has to be more than this". Then you started reading books and magazines about far off places and watching movies that demonstrate achievement and the power of change, which fueled your inquisition. Then over time your theory turned to passion and then into action and you left home, taking your drive and intolerance for status quo with you. Following the advice of Horace Greeley...... "Go west young man and grow up with the country"

This process happened to millions of people and is still happening every day. The frontier countries of Australia, Canada and the United States are full of these people, which has made them what they are today, thriving nations of 'never say no' go getters. 'Impossible' to these people is just 'possible' that hasn't been done yet. This is the true reason why emigration and immigration are so important to growth, look around you at the people in your community who have started with nothing and built businesses that are the street level heart of your town. They are hardworking people from somewhere else with a predisposition for continual growth.

So picture this: you take a country full of these types, (Australia, Canada, USA) then you pick an industry that is all about frontiers (Construction) and you find some place that is just past the end of paved roads (Darling Downs, Fort McMurray, Williston North Dakota) and you fill it with these people. First thing you notice is that it is hard to get a coffee or a meal because everybody wants to change the world and you can't do that in the service industry. Then of course you can't get anywhere on the roads because the infrastructure is way behind and when you do get two lanes everybody is in the fast lane. But what you do end up with is the ultimate mixing bowl of drive, testosterone and a need for accomplishment, the ultimate recipe for a revolution.

Then we add a series of mega projects and a stagnant process for construction execution that is based upon an old model for Engineering then Procurement and then Construction, rather than the parallel execution of Fast track construction, and voila we have the formation of a fault line.

And this is my theory for where the spark for WFP, IM and AWP came from. It started in Fort McMurray with this churning caldron of old processes and frontier people and then spread to other frontiers that had the same mix of frustration for improvement. I've seen the light come on in construction people who had lost all hope of ever doing things that made sense. We would roll out WFP to them and you would see them cautiously sniffing around it, kicking the tires to see if it was real or just a cardboard cut-out. At that moment when folks realized that they could shape their own outcomes, all that drive that sprung them from their home town would come back to life. Even now when we roll out WFP on projects where all hope appears to have been lost, we see that it becomes a magnet that attracts the doers.

The fifth element in the factors of productivity is 'Desire':

Information, Tools, Materials, Access and Desire. We normally don't even mention this when we are setting up targets for projects because we know that 'desire' is only dormant in most people, just waiting for the opportunity to bubble to the surface.

To Summarize 'Why Change?'

We covered the reality and pace of change in the world around us then we spent some time looking at the motivators for change in our industry and identified the Owner as the primary 'Change Agent'. This led to a quick review of our current reality and our new desired state. Then the last couple of pages showed that we have the ability and desire for improvement.

The next discernable set of questions are: What do we change? and how do we change it?

The short answer for 'What' is: Implement systems that address the issues of Schedule and Cost uncertainty

The short answer for 'How' is: Apply Advanced Work Packaging because your Peers have and it worked for them.

Chapter 2: What is

Advanced Work Packaging (AWP)
Information Management (IM)
Workface Planning (WFP)

In the introduction, we used the CII surfboards to map out the high-level structure of AWP, IM and WFP. In this chapter, we will go one step deeper and explore the actual components of each area and discuss the flow between them. Then in the following chapters we will get into the weeds of who does what.

If you haven't already viewed the AWP infographic from You Tube or our website (www.insight-awp.com) this would be a good time to do it. The infographic video will take you through the flow chart and give you a high-level review of how all the parts contribute to the final product.

Advanced Work Packaging:

The simple explanation is that the process of Advanced Work Packaging guides the dissection of project scope so that it supports the execution of Workface Planning in the field.

It starts with the processes upstream of the Construction Work Package and aligns engineering work packages with procurement work packages. This populates the construction work packages with all the drawings and materials and gets them ready to be carved into Installation work packages by the construction team.

The other two essential elements of the bigger picture are: Information Management and Workface Planning:

Information Management:

The management of information is a strategy that starts with the idea that everybody on the project needs information that is created by somebody else on the project. Therefore, the target of the strategy is to design systems and interfaces that align the source data with the users. The desired outcome is to ensure that all the right people have access to the data that they need, that it is compatible with other project data and it is formatted to be interoperable.

So that everybody knows everything that they need to.

Workface Planning:

In a coal mine the workface is that point where the pick hits the coal, in our world of construction it is that point where trades people turn materials into a functioning plant. Therefore, Workface Planning is the process of identifying what these people need and what we must do to get it to them.

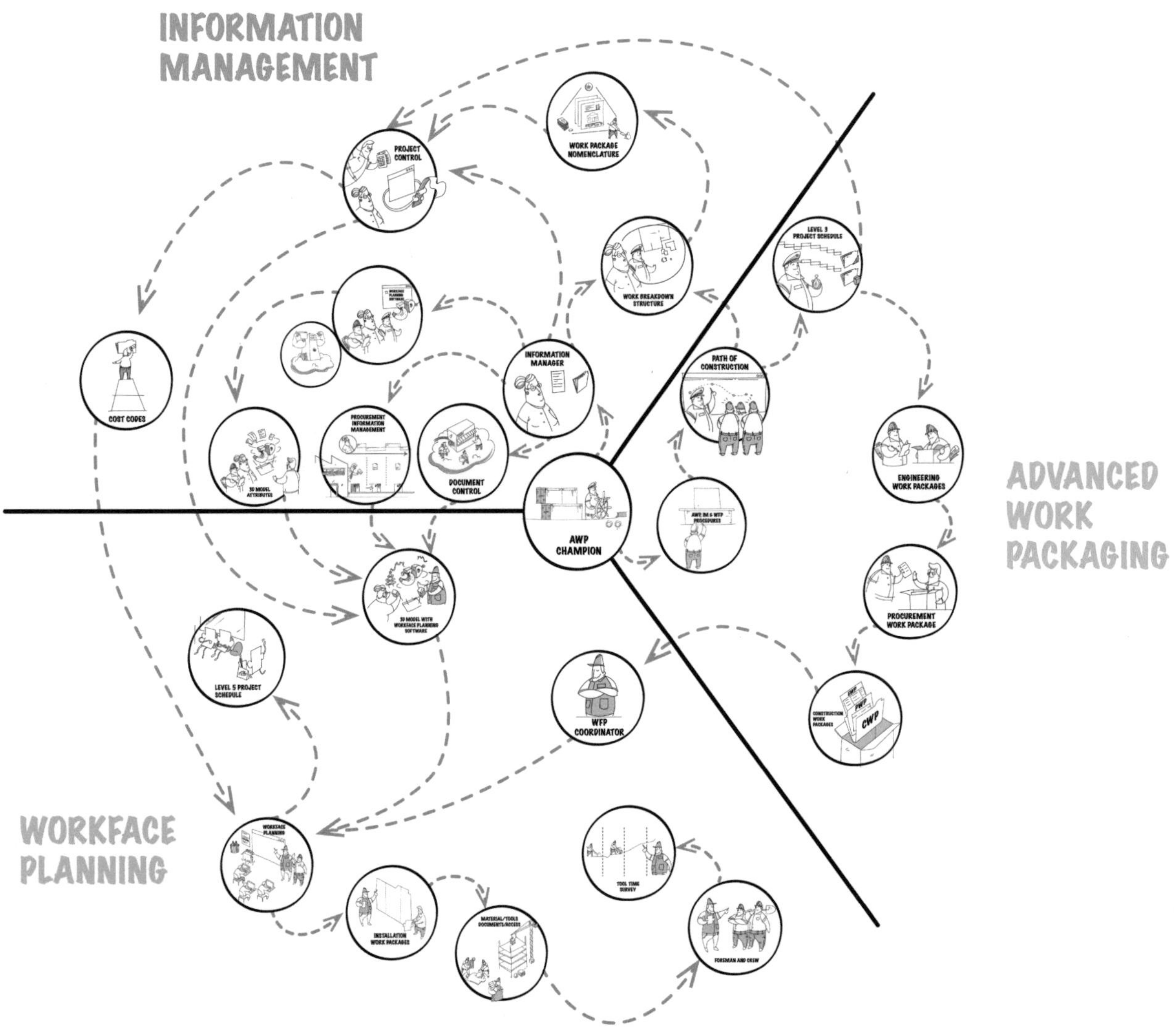

As you can see in the infographic, the book and the procedures are separated into three distinct areas, which represent three overlapping phases of the project.

Advanced Work Packaging starts before front end engineering with the assignment of the AWP Champion, who is charged with leading the department, creating the overall strategy and developing an overall culture of work packaging.

The Information Manager is appointed just a short time after that in time to influence the development of rules and standards for information generation and exchange.

The third phase of the process is Workface Planning which is initiated by the appointment of the Workface Planning Coordinator at the start of detailed engineering design.

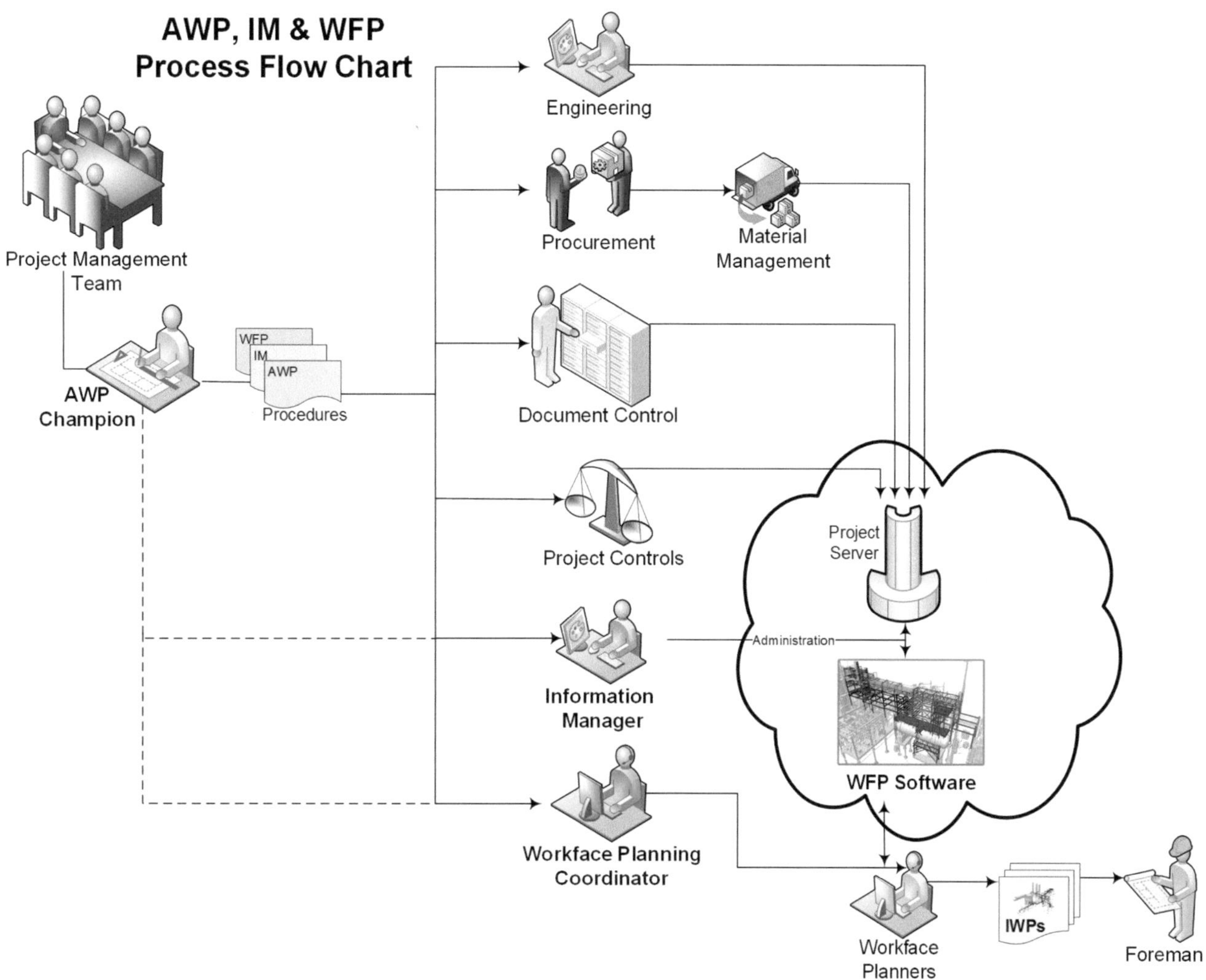

The procedures detail the roles and responsibilities of the team, while also establishing the overall expectations with other stakeholders.

Chapter 3: Return on Investment & Benefits

If you read through the literature from The Construction Owner's Association of Alberta (COAA) or the Construction Industry Institute (CII) you will see that they have both identified that you can reduce the Total Installed Cost (TIC) of projects by 10% by applying AWP. This is a result of improving the productivity of direct labor by 25%.

Sounds like big numbers and a little bold to be throwing out 'pie in the sky' figures like this, but the truth is that if they published the real numbers then absolutely nobody would believe them. Which is both exciting and a little sad. It means that the average project is so far in the ditch right now that any measure of organisation can flip a 25% productivity improvement. The exciting part is that there is lots on the table.

To give you an idea of what is out there we recently conducted a Time on Tools study between two areas on the same brownfield project: One area was in a shutdown – very well planned and packaged, the other was not planned and was being executed 'De la Journee' (of the day). The shutdown area had direct activity of 47.37 % which is amazing by itself and the De la Journee area had direct activity level of 25.6% (47.37 – 25.6 = 21.77, $\frac{21.77}{25.6}$ = **85%**). Sounds incredible (not credible) and you are right, but just stop for a moment and think about running a shutdown with just high-level planning. You walk into the plant one day with a broad scope of work, shutdown the facility and then send crews into the field to find and execute 'opportune work fronts' that you don't have materials or documents for. Do you think that the shutdown could take 18.5 days instead of 10? Absolutely, and that's what we are doing on our standard construction projects.

Having said that I don't think that we are ready to start planning our two-year construction projects at level 6 (daily) or level 7 (hourly), but we certainly can plan and execute rolling wave level 5 (weekly) schedules predictably.

To show you that there is lots on the table and that this is not a 'one off' outlier here are some other results that we have recorded on projects over the last few years.

Project A: Canada, 2006, $300 Million, WFP software applied with IWPs for all foremen in the field, 3rd party audit of WFP scored 77%

The other half of the project was a different contractor not using WFP

WFP side: Installed Pipe at 2.6 hours per foot

Non-WFP side: Installed pipe at 3.4 hours per foot

= 24% delta

Scaffold cost was 18% of direct labor (normally 25%)

Project team were awarded the COAA award for the 'Best Implementation of a Best Practice' 2007.

Project B: Canada, 2008, 15-day 500 Craft shutdown, Job Cards for every foreman, every day: Shutdown finished 1 day early with significant added scope and a perfect safety record.

Project C: Canada, 2010, $500 Million, series of sustaining projects over two years, 5 contractors, IWPs for each foreman each week:

One company went from highest safety incidents in class to lowest safety incidents in class after adopting IWPs. Another company logged a significant increase in profits and transitioned to become the best performing division of their company.

Project team was awarded the COAA award for the 'Best Implementation of a Best Practice' 2011.

Project D: Canada, 2012 $1 Billion, 400 Craft, 126-day shutdown, Job Cards for every Foreman, every day: Finished on time with significant extra scope, perfect safety record and **13%** under budget.

Project E: Canada, 2013, $1Billion, 500 Craft, 3 years, IWPs adopted mid-stream, WFP software applied:

Identical scope for train 1 and 2 on a 20,000-hour CWP, one with IWPs and one without:

The CWP with IWPs built from WFP software recorded a 32% increase in productivity over the CWP that was constructed De la Journee.

Project F: Australia, 2013, $500 Million, yearly cycle of 500 Craft shutdowns on a series of mine sites. IWP and Job Card format adopted. Constraints tracked in Access database:

Planning Team recorded a full year of successful shutdowns, ahead of schedule or on time with the worst performing shutdown coming in 12 hours late.

Project G: USA, 2014, $400 Million, 500 Craft project, WFP software applied, IWPs for every Foreman:

Project recorded a $36 Million cost avoidance, perfect safety record, scaffold cost was 18% of direct labor, pipe was installed at 2.6 hours per foot.

Project team was awarded the 'Be Inspired' award for the best global application of WFP software 2016.

Project H: Canada, 2015 $100 Million series of sustaining projects, Job Cards or IWPs applied for every foreman.

Construction company transitioned from the least preferred (mandated) contractor to the contractor of choice after a series of 'under budget, ahead of schedule' projects.

Project I: Canada, 2015, $5 Billion project, Multiple contractors with a wide range of WFP compliance, 5000 Craft workers, WFP software applied in pockets:

Tool time studies and WFP audits across all contractors showed a 17% difference in productivity between contractors using the process well and those not using WFP very well. This worked out to a $71 Million cost reduction and 65 days less of schedule for the areas using WFP well.

Project J: Europe, 2016, $500 Million project, WFP adopted midstream on half of the project areas:

Tool time studies conducted every two months showed a 31% difference in the average direct activity levels (productivity) of WFP areas over non-WFP areas.

This is only 10 of the more than 30 projects that we have been involved with so far. The rest of the industry have many more examples of the positive impact of WFP and then the extended benefit of AWP. The key for you to know is that right now construction projects have the option to significantly impact their productivity, schedule, predictability and safety by applying AWP. Sometime soon you may not have the choice. If the evolution of productivity management follows in the footsteps of Safety and Quality management, then project managers will have to explain why they didn't apply this proven best practice.

The same development cycle is common amongst other industries that go through transformative growth. There are early adopters, who see the benefits of change and get on board for the right reasons and there are those who refuse to see the need for change, who get caught in the tide against their will: Mammals and Dinosaurs. Your choice is to decide which group you will be with.

Either way the results that we are seeing now are not a flash in the pan from the productivity initiative of the month. They are the outcome of many years of hard work, research, development and the application of lessons learned, that have created a model for project execution that delivers predictable outcomes. Several of the companies that we have worked with have already reached that tipping point where they realize that this is a business model for sustainability, not a 'nice to have' option or a sales pitch.

The Investment:

First of all, let's try to keep things in perspective because the numbers we are going to explore get way out of whack when we look at AWP as an investment.

I would like you to think about the sort of Return on Investment (ROI) that you might feel comfortable with for your retirement investments. For myself I think of blue chip investments that have a steady return of 5-10% as being predictable and that is where I place most of my investment dollars. As a risk taker, I am also obliged to have some foreign start up bank/Ponzi scheme/loan sharking/Antarctic gold mine investments that promise returns in the 25-50% range. Which I think of as like betting on a single number on the roulette wheel, it's probably gone, but it's exciting to think that it could come in.

Our most recent project to be completed had an upfront investment of $3Million (Planners, software and administration) and a quantified return of $36 Million. That is an **1100% ROI.**

Plus: a perfect safety record, lots of peaceful night's sleep for the project team and a good helping of career boosting project performance that will last at least 10 years.

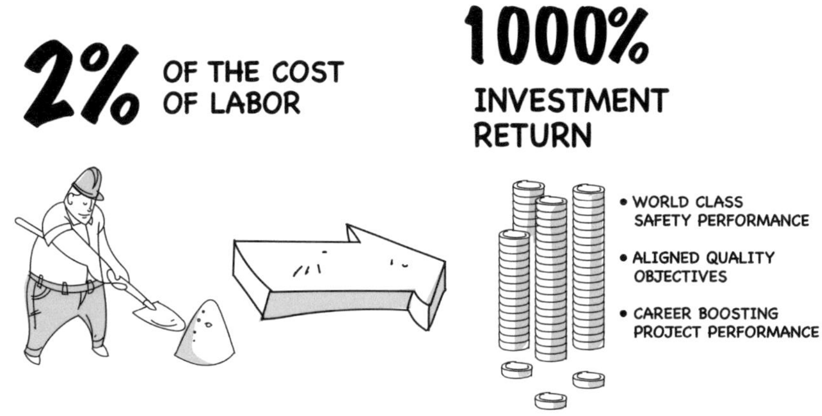

If an investment banker had a stock that performed like this with a steady increase in predictability tracked over the last ten years, there would be a stampede of investors. The good news is that this investment is not limited by a definitive number of shares or subject to a price increase once the secret is out. There is an infinite supply of cost reduction opportunities out there, you just have to decide when to invest.

Predictability:

Enhanced performance and the probable cost and schedule reductions that come from AWP create a new problem for us: One of our obligations in project management is to be predictable, specifically in the fields of schedule and cost, so while it is nice to finish a project early and under budget, it is even better to have predicted that outcome so that we didn't tie up funds and schedule. The ideal project finishes right on time and that enables other stakeholders to take advantage of reliable turnover and start up dates.

So how much do we reduce our cost and schedule estimates when we are applying AWP? The safe answer is to stick with the proven 25% productivity increase recorded and shared by COAA and CII and apply this reduction to the standard for commodity installation rates. Traditionally the target of WFP was only steel and pipe but our history has shown us that when AWP and WFP are applied to all disciplines it has a similar impact on productivity levels across the board. As an example, we now know that the application of AWP in Nth America will deliver a pipe installation rate of **2.6 hours per foot,** down from 3.5 or even 4. A rate that we have seen achieved several times on projects over the last few years. The same level of productivity increase has been experienced with steel, E&I and civil disciplines, which means that you should probably apply the 25% increase to all disciplines.

However, the entire question is a little more complex than just slashing 25% from your budgets, there is an investment that you need to make to get this return. There is the obvious cost of Administration, Workface Planners and software, (2% of your direct budget) but there is also some added cost and schedule in engineering and fabrication that is required to deliver packages not just drawings or spools. It is best assessed on a case by case basis but it is safe to assume that your engineering and fabrication cost and schedule will increase by up to 5%.

So as an overall summary of your investment I will us this WAG (Wild Ass Guess) model:

On $1Billion, 3-year project:

Total Installed Cost = 10% Engineering, 50% Procurement, 40% Labor.

Engineering $100 Million plus $5 million, 18 months plus 20 days

Procurement $ 500 Million plus $25 Million, 18 months plus 20 days

Labor $400 Million minus $100 million, 2 years minus 130 days.

= a reduction of $70 Million and 90 days (plus the reduction of overheads and administration for documents, materials and project controls), which is a little short of the 10% TIC that we talked about but it is still within the realm of possibility.

Having driven this stake into the ground for targeted cost and schedule reduction, you have also established priorities, goals and a global vision of the project for your team. This holistic, 'construction first' strategy replaces the 'silo first' model that has cost us dearly over the last couple of decades. It's certainly true that the optimization of engineering and procurement is often at the expense of construction. This model should help you prevent that behavior.

Chapter 4: AWP Quick Start Guide

I'm guessing that the last time you purchased electronics or 'assemble at home' furniture that there was a quick start guide with big pictures and simple instructions that came with it. This chapter is the AWP equivalent of that guide. If you just want to get into action and get a system up and running without becoming the regional expert, this is your section. It is also a good place to start if you want to know the difference between 'critical elements' and 'nice to have enhancements, which is the key to building a scalable model. The industry has applied this process to lots of projects now and we understand that if you only have Workface Planners, Installation Work Packages and some form of Constraint Management (The critical stuff), you will have the basic model for Workface Planning and get lots of benefit. This quick start guide goes one step further and identifies the critical components for Advanced Work Packaging. Much like new electronics or software that you have running in the basic mode, there are also lots more features! If you do go on and read the details in the other chapters you will discover the enhanced model is pretty slick too.

Quick Start Guide: The aim of Advanced Work Packaging is to ultimately give each Foreman an Installation Work Package (IWP) at the start of each week that is 'ready to go': The scope is identified, the material and tools are available, the prerequisite work is complete and the scaffold is erected and fit for purpose.

The expectation from the Superintendent is then that the foreman and crew will get the work done within the scheduled window of time and estimated hours. The results are typically a 25% increase in productivity and a 10% reduction in the Total Installed Cost of the project.

There are many streams of contribution that lead to this state and this Quick Start Guide is designed to identify the core components that must be in place in order to effectively apply Advanced Work Packaging (AWP), Information Management (IM) and Workface Planning (WFP)

A. Advanced Work Packaging Champion
B. Advanced Work Packaging Procedures
C. 3D Model and Fabrication Data
D. Workface Planning Software
E. Workface Planners
F. Installation Work Packages
G. Constraint Removal
H. Project Controls
I. Field Execution

A. Advanced Work Packaging Champion.

The application of AWP needs direction and guidance, much like the way that Project Management organizations apply models for Safety and Quality. The assignment of a dedicated Champion within the Project Management organisation is a critical first step to the development of a culture of productive project execution.

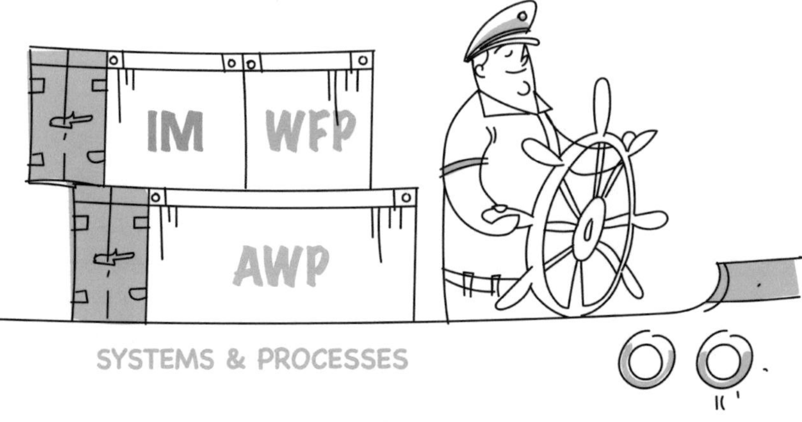

AWP CHAMPION

As a member of the Project Management team, the AWP Champion typically represents the Owner and coordinates deliverables from all of the project stakeholders impartially, with the singular focus of bringing benefit to the entire project.

The AWP Champion functions as the project coach, guiding the implementation of AWP, IM and WFP procedures. Aligning engineering and procurement with the needs of construction, developing IM standards and processes and then the development and application of Workface Planning in the field.

The ideal candidate for this position is a Subject Matter Expert in the field of AWP and Project Management, with experience in Construction Management.

B. Advanced Work Packaging Procedures

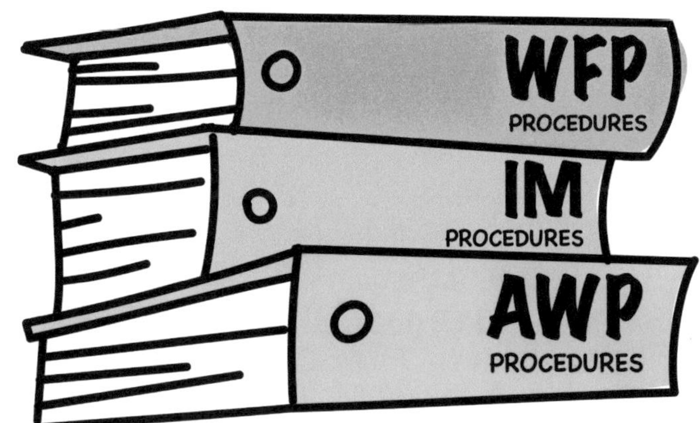

- Workface Planning
- Information Management
- Advanced Work Packaging

In order for the project stakeholders to follow a set direction there must be a standard to go by. While procedures alone do not make changes happen, they do establish an expectation for compliance and will facilitate audits of the process later in the project. Ideally procedures are specific, detailed and identify: Who, What, When, How and Why, with flow charts and templates.

Importantly the procedures need to address three key areas:

Advanced Work Packaging: Contract language, the optimal path of construction, the sequence of engineering & procurement work packages and standards for project controls.

Information Management: Standards for the creation of the WBS, project nomenclatures, data development, the application of WFP software, software interfaces and the structure & attributes of the 3D model.

Workface Planning: Workface Planners, installation work packages, constraint management, field execution and project reporting.

C. 3D Model and Fabrication Data

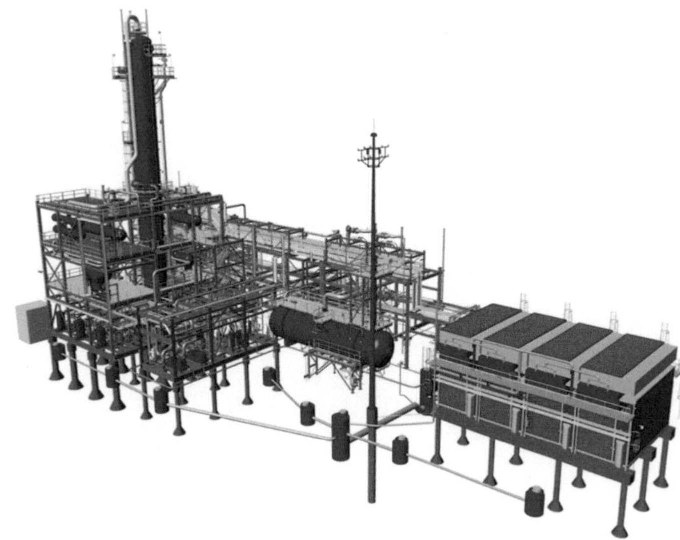

- Isometrics
- Cut sheets
- Erection drawings
- Cable schedule
- IDF, PCF and CIS2 files

The creation of a fully attributed 3D model and intelligent data created by the fabricators is already produced on most projects. Obtaining this data and making it available to the WFP team facilitates the planning process and creates a solid platform for the communication of project information.

The process that leads to the delivery of this data is typically the inclusion of a contract clause that identifies the data as a deliverable for engineering and fabrication. The project management team then establish a delivery schedule that ensures that the project data is complete and kept up to date.

Typically, the AWP Champion and the Project Management Team facilitate the development of a matrix of attributes for engineering to guide the population of the 3D model during design. This will allow the construction team to mine the appropriate data during the construction phase.

D. Workface Planning Software

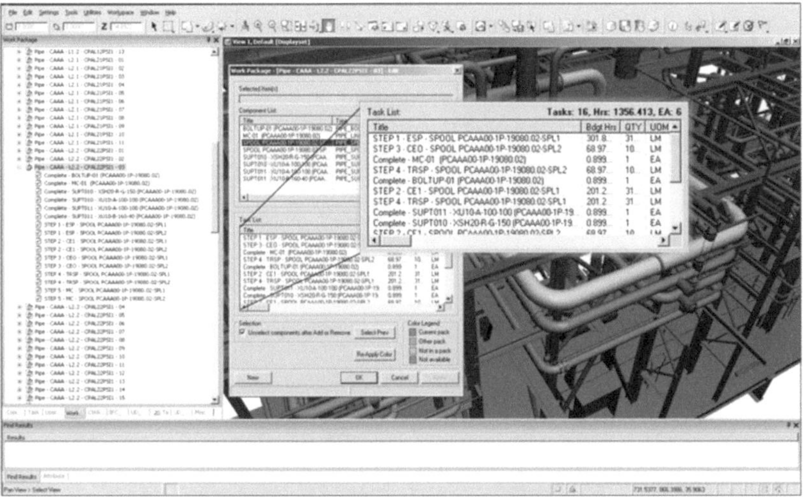

WFP software is commercially available software that organizes the project data on the platform of the 3D model so that Workface Planners can develop and manage Installation Work Packages (IWPs) in a virtual 3D environment.

The software also facilitates the calculation of Planned Value by hosting installation unit rates and rules of credit. It can also be used to create 4D simulations of IWPs based upon the project schedule.

Prior to the start of FEED, the AWP Champion and the Information Manager assess the features and compatibility of the different products to find the best fit with their existing software. This would typically span the software used for the 3D model, material management and document control.

E. Workface Planners

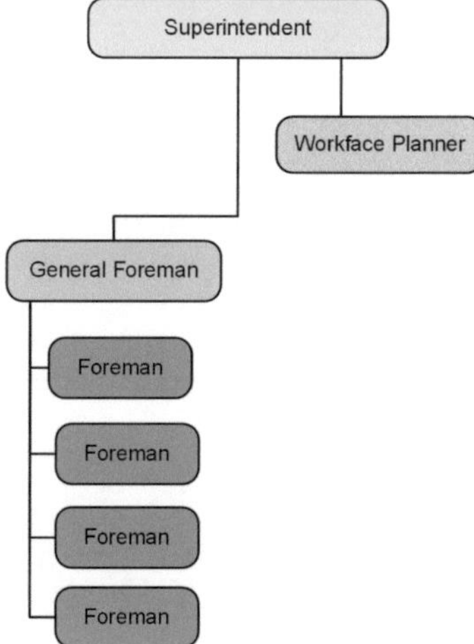

The critical component of the Workface Planning process is the creation of an extra position within the typical construction organisation.

The Workface Planner is dedicated to developing plans for the Foreman based upon the execution strategy developed by the Superintendent.

A typical profile for a Workface Planner:

- ✓ Tradesperson who understands the work
- ✓ Experience in field supervision and execution
- ✓ Basic computer skills
- ✓ Will be a member of the construction organization
- ✓ Will report directly to the Superintendent
- ✓ Each Superintendent will have one

Once you have identified the candidates enroll them into a Workface Planning Training course. (Along with the Foremen and Superintendents so that they understand the context)

Give each Workface Planner a copy of this book to read prior to starting work.

Set up a common area with desks, computers and phones for each Workface Planner.

The approximate ratio of Planners to Field Workers should be **1 to 50** with some consideration for complexity. The complexity of instrumentation requires more planning resources than pouring concrete or earthworks.

F. Installation Work Packages

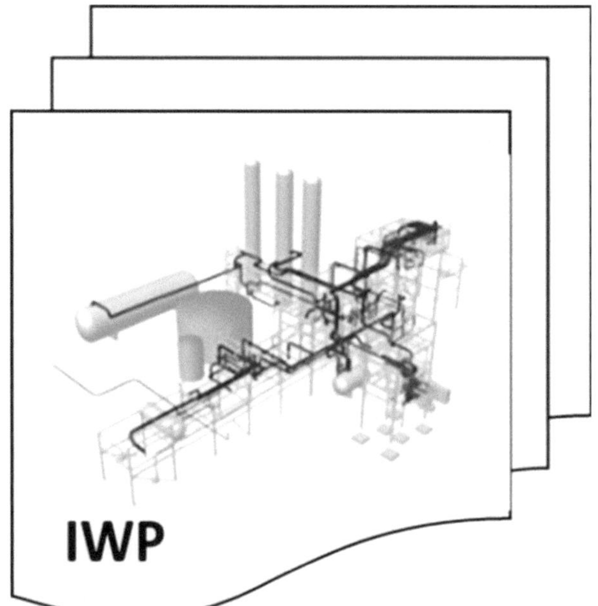

The model for Installation Work Packages is based upon this simple question:
What does a Foreman need to execute work?
The answer then becomes the contents of the IWP.

The generic standard for a single discipline Installation Work Package:
Cover page: IWP name, Planned Value and a 3D of the scope.
Emergency Information: Names and phone numbers.
Constraints: List of elements to be satisfied or removed.
Scope statement: Basic sequence and scope statement
Safety: Requirements specific to the scope
Quality: ITP requirements for signatures and inspections
Documents: Complete list and copies of technical documents
Materials: Complete list of the materials required and confirmation that they are available.
Access: Permits, scaffolds, pre-requisite work and trade coordination
Equipment: Detailed requirements for specialty tools and equipment
Timesheets: Copies of timesheets and a list of the correct cost codes for the scope
Delay codes: A list of delay codes and instructions on how to record them
Progress: A progress matrix that shows components, stages of completion and Planned Value.
Schedule: A snapshot of the 3 Week Look Ahead showings this IWP and dependencies.
Completion: Allows the foreman to record incomplete work and lessons learned.

Size and Content: How much work should be in an IWP?

Start with one rotation, approximately 500 hours and then allow the model to develop into a fit for purpose application. Generally speaking, smaller packages are better for the Foremen, they are easier to track & close and will guide the execution of a specific sequence.

Developing IWPs: Pick a CWP that is planned for execution in 90 days and ask the Superintendent to sit with the Planner and describe how the work should be dissected and sequenced.

The Planner develops the IWPs in the WFP software 3D environment and drafts a scope statement for each one, then the Superintendent reviews and approves the dissection and sequence. The Planner then builds each IWP and populates each of the sections based upon the specific scope.

G. Constraint Removal:

The process of constraint removal and management is the single biggest change from the way that construction is typically executed. The golden rule is that IWP's must not be released to the field until they are free of constraints and are 'ready to execute'.

Critical constraints on an IWP are typically **Scope, Documents, Materials and a Technical Review that identifies any RFIs.** When these issues are satisfied the IWP joins the backlog, where it sits until it enters the Three Week Look Ahead. This triggers the actions to address the secondary constraints: scaffold, construction equipment, safety, quality, workforce and preceding work.

This process is managed by the Workface Planners in Pack Track (below), which is stored on the project cloud and shared with the project as a hard copy in the War Room.

As the constraints are cleared each IWP passes through the gates and becomes available for execution.

				90 Day Planning			IWP Assembly				3 Week Look Ahead										
		Weeks prior to execution		12	12	12	4	4	4	4	3	3	3	3	3	2	2	2	1	1	-1
CWP PE3-57	IWP	Description	Planned Value	Scoped	IWP Created in 3D	Inserted into L5 Schedule	Documents IFC	Materials Available	Technical Review (RFIs)	Enter Backlog	Enter 3 Week Look Ahead	Bag and Tag Material	Request Scaffold	Request Cranes & Equipment	IWP Hard Copy	Safety	Quality	Resources Confirmed	Preceding Work Confirmed	Issued to the Field	Work Complete
Civil																					
PE3-57-EW																					
Grade	PE3-57-EW-01	Survey for Grade	840	✓	✓	✓	✓	✓	✓	✓	✓	✓	✓	✓	✓	✓	✓	✓	✓	✓	✓
	PE3-57-EW-02	Strip Top Soil	1340	✓	✓	✓	✓	✓	✓	✓	✓	✓	✓	✓	✓	✓	✓	✓	✓	✓	
	PE3-57-EW-03	Grade to Elevation 1	890	✓	✓	✓	✓	✓	✓	✓	✓	✓	✓	✓	✓	✓	✓	✓			
	PE3-57-EW-04	Grade to Elevation 2	730	✓	✓	✓	✓	✓	✓	✓	✓	✓	✓	✓	✓	✓	✓				
Piling	PE3-57-EW-05	Survey for Piling Placement	620	✓	✓	✓	✓	✓	✓	✓	✓	✓	✓	✓							
	PE3-57-EW-06	Mobilize Piling rig and materials	450	✓	✓	✓	✓	✓	✓	✓	✓	✓	✓								
	PE3-57-EW-07	Install Piles North Side	980	✓	✓	✓	✓	✓	✓	✓	✓	✓	✓								
	PE3-57-EW-08	Install Piles South Side	730	✓	✓	✓	✓	✓	✓	✓	✓										
	PE3-57-EW-09	Cut and Cap Piles North	860	✓	✓	✓	✓	✓	✓	✓	✓										
	PE3-57-EW-10	Cut and Cap Piles South	1250	✓	✓	✓	✓	✓	✓	✓	✓										
PE3-57-CO	PE3-57-CO-01	Survey for form work	820	✓	✓	✓	✓	✓	✓	✓	✓										
Formwork	PE3-57-CO-02	Excavate for form work	1420	✓	✓	✓	✓	✓	✓	✓	✓										
	PE3-57-CO-03	Install form for EB-43	850	✓	✓	✓	✓	✓	✓	✓	✓										
	PE3-57-CO-04	Build Rebar cage EB-43	640	✓	✓	✓	✓	✓	✓	✓	✓										
	PE3-57-CO-05	Construct Forms for CG3-9	790	✓	✓	✓	✓	✓	✓	✓	✓										
Rebar	PE3-57-CO-06	Build Rebar cage CG3-9	550	✓	✓	✓	✓	✓	✓	✓	✓										
	PE3-57-CO-07	Pour EB-43 and CG3-9	350	✓	✓	✓	✓	✓	✓	✓	✓										

H. Project Controls

Project Schedule: As a CWP enters the 90 Day window the Workface Planners will convene a meeting with their Superintendent and dissect the CWP (Level 3 activity) into IWPs (level 5 activities). The Workface Planner arranges the IWPs into a sequence based upon the Superintendent's execution strategy and submits the arrangement to the Scheduler as the IWP release plan. The level 5 schedule should then only contain IWPs that roll up to CWPs and milestones. Typically, we leave the option for level 4 activities out of the schedule, but available if cost management requires that they be tracked.

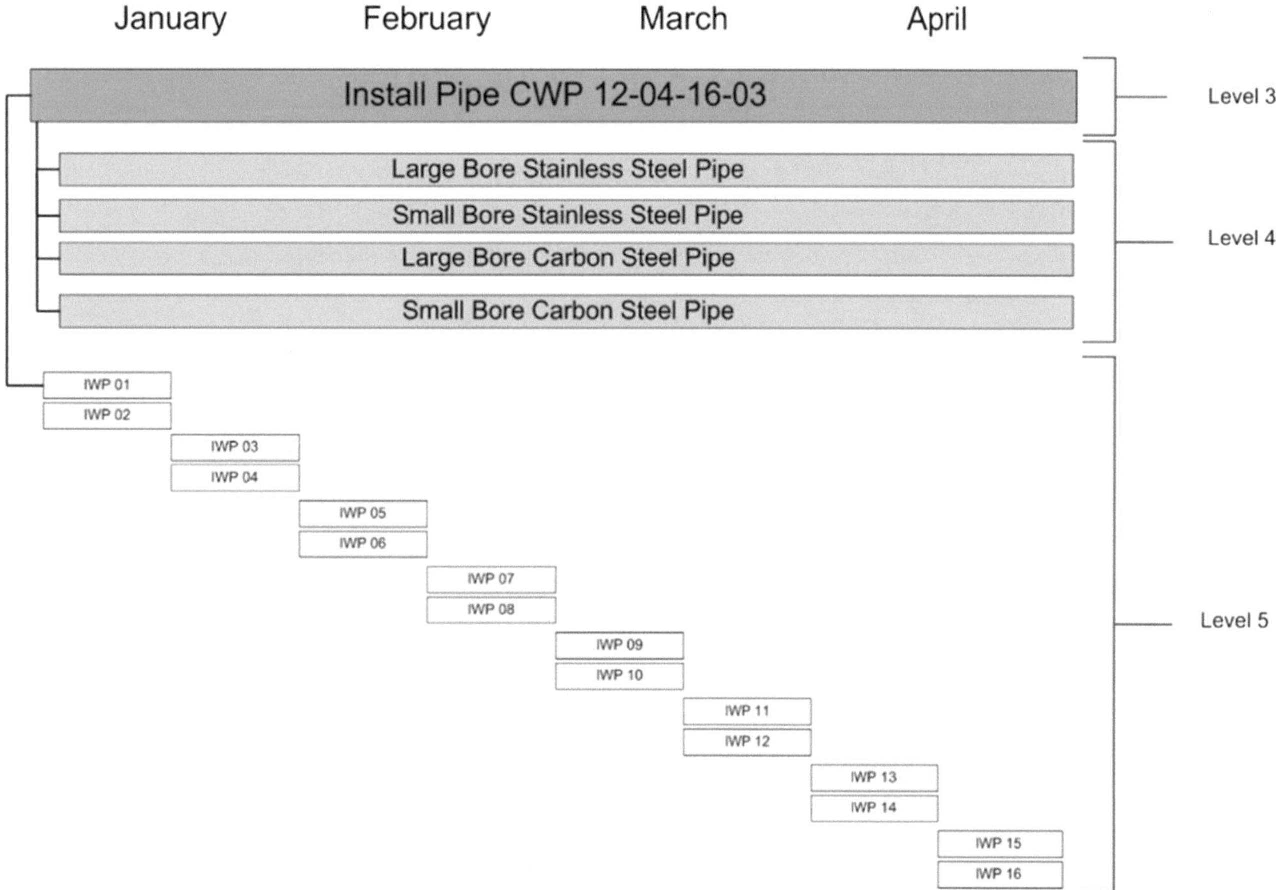

Earned Value Management: The identification and tracking of Planned Value against Earned Value is one of those things that you will wish that you had done if you don't. It is the foundation for understanding the critical answers to: How much is done? How much do we have to go? How much will it cost? And How long will it take? The WFP software will help you calculate PV for steel and pipe, everything else will need to come from your Project Controls department, so let them know that you will be calling. PV must be on the front cover of every IWP.

Three Week Look Ahead: Working from the level 5 construction schedule, the Superintendents create a three week look ahead that is updated each week with constraint free (yellow) IWPs from the backlog. This triggers the Workface Planner to initiate requests for materials to be bagged and tagged and scaffold to be erected. The Construction Management Team then combine all of the discipline specific three week look aheads to create a single integrated three week look ahead for the whole project.

I. Field Execution

The culmination of these efforts results in IWPs that are constraint free and "ready to go', however that is not the end of the line. As victims of their environment, you will find that most Field Supervisors are used to living hand to mouth. If you want them to transition to trusting a plan, then you will have to earn that trust and coach them on how to think in one week increments instead of daily blocks.

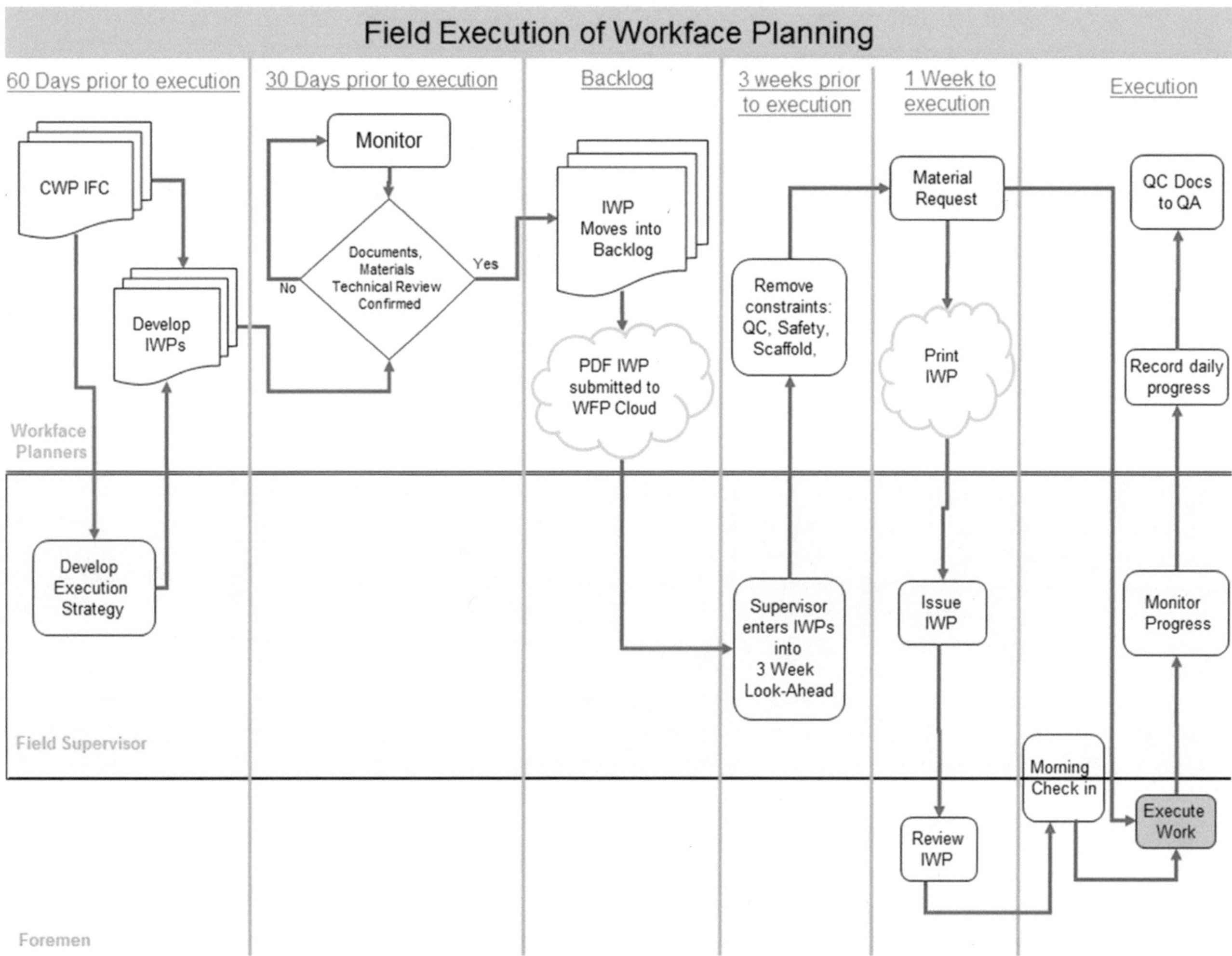

This flow chart shows the hard copy IWP being printed and issued to the Foremen in the week prior to execution. This allows the Foreman some preparation time which starts the process of task organisation that will lead to completion of the scope within one week.

The next vital step is to institute a 15-minute morning Check in Meeting, which takes place before the crews arrive. The Foremen meet with their Superintendent and lay out a level 6 plan for their day on a whiteboard. Then each subsequent morning the Foremen update their progress against the plan. The Superintendent monitors their progress and coaches them towards completion of the IWPs within their one-week window.

It sounds simple and it is true that good Superintendents already do this, but it is something that you cannot leave to chance, it would be a shame to drop the ball here after all the effort of generating the IWPs.

Importantly each IWP has a use-by date which is normally at the end of the week. The IWPs must then be returned so that the Planner can extract any remaining scope and move it into a punch (clean-up) package. The completion and return of the IWPs is then the trigger for QC verification, which ensures that the standards have been met and that the work is done, done.

Chapter 5: Advanced Work Packaging

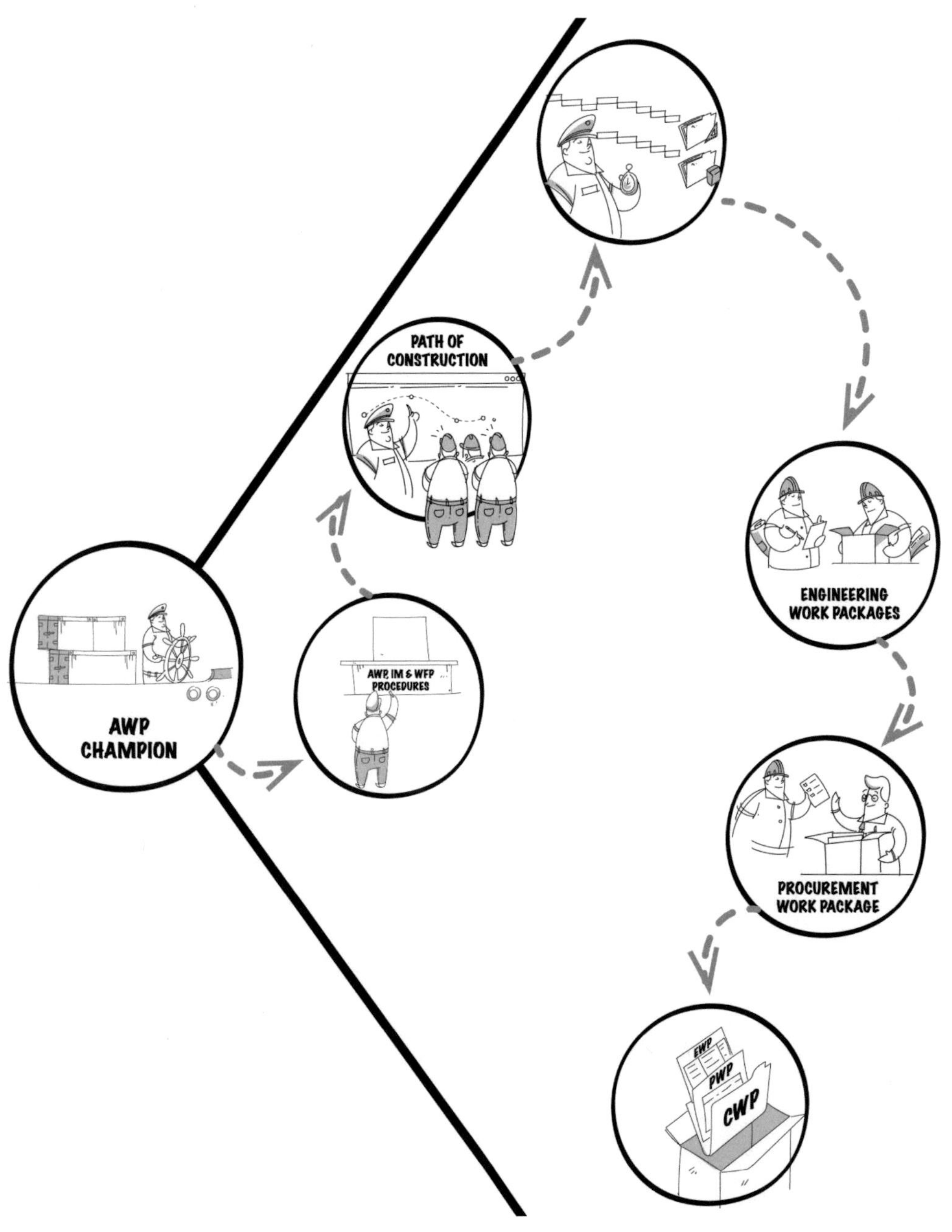

ADVANCED WORK PACKAGING

A. AWP Champion
B. Information Manager
C. Procedures
D. Work Breakdown Structure
E. Path of Construction
F. Level 3 schedule
G. Work packages for Engineering, Procurement. Modules and Construction

Advanced: progressive, forward thinking, unconventional, cutting edge, innovative, radical

Work: Composition, design, creation, opus, masterpiece, product, handiwork, oeuvre

Packaging: Wrapping, parceling, boxing, bundling, enveloping, inclosing, tie together, presenting.

As discussed earlier in the book the term Advanced Work Packaging is a bit of a play on words that suggests that the process is both forward thinking and at the beginning of a project. Developed by the CII Research Team 272, the term is an apt description of the preliminary ground work that needs to take place at the start of a project to align engineering deliverables with the path of construction. Having said that I also think that it could have a sub title that would then describe what it means.

- Project execution by work packages
- How to align E & P with C
- The foundation for Workface Planning
- Fast track (parallel) execution in series
- The language of work packages
- Work packages for fast track projects.
- The rules of work packaging.
- How to align Engineering with Procurement so that Construction can build in sequence, be productive, finish on time and cost the least amount, while also optimizing safety and quality.

The skinny on AWP is that it is the ground work that establishes a model for Construction to receive drawings, modules and materials, so that they can effectively implement WFP.

The logic starts with the project's commitment to AWP through the assignment of an AWP Champion. It is this person's task to establish procedures for AWP, IM & WFP and to appoint an Information Manager who leads the development of the Work Breakdown Structure. The WBS logic that aligns Engineering, Procurement and Construction work packages is then one of the key inputs for the Path of Construction workshops. The output from the PoC is the outline of the level 3 schedule made up of EWPs, PWPs, CWPs and major equipment, which creates the foundation for a culture based upon work packages.

Along with these developments, the Champion is also responsible to contour behaviors so that they support the efficient implementation of these processes.

Sounds easy if you say it fast.

To give you an order of magnitude, on a one-billion dollar project this process will likely take 6 months to establish if things go well....... and then daily management through the lifecycle of the project.

A. AWP Champion

The assignment of an AWP champion establishes the project's commitment to the application of AWP and initiates the process of development and execution that will guide the project through to completion.

It is important to note that the successful execution of AWP is based on the premise that there is a true Project Management Team, (PMT) typically made up of Owners, 3rd party SMEs and representatives from the E, P & C organizations. It is the collective vision of holistic project success that gives an autonomous PMT the correct platform to administer AWP. This becomes very apparent when we introduce systems that augment construction at the expense of engineering or procurement.

The Champion: I have had some resistance to the title of 'Champion' and you may find a better term but the meaning is not that the person is the very best at AWP (noun) but rather that this is the person who is ultimately accountable to champion (verb) the process. The go to person for all things AWP.

Ideally the Champion will report to an AWP Sponsor who is an executive member of the Project Management Organization. This establishes the authority for them to operate and the expectation that the project will be AWP compliant.

To mix a couple of quotes from Eric Crivella and Yogi Bera: "The application of AWP is 90% sociology and the other half is physical." So, your Champion needs to know how the whole thing works from personal experience, have a fanatical focus on 'construction first' and be a hardnosed, sociably amiable person who can influence the behavior of others.

B. Information Manager

The assignment of the Information Manager is one of the first tasks that the AWP Champion will address. A full description of the profile and the process is covered in the next chapter.

C. AWP, IM and WFP Procedures

One of the early tasks that the Champion will tackle is the development and roll out of an AWP execution plan that lists all the major milestones, a sequence of events and the factors of success. The cornerstone of this execution plan will be the creation of procedures for AWP, IM and WFP. This process can be facilitated by the engagement of an Information Manager and a Workface Planning Manager, whom both report directly to the AWP Champion.

Even with this team established the creation of procedures can be a daunting task, if you must start with a blank page, however there are resources out there that can help. Both COAA and CII have lots of samples on their websites and there are 3rd party generic procedures available from several of the industry service providers. The key is to end up with a set of 'fit for purpose' procedures that identify who does what and when, what the deliverables look like, a training plan and how you are going to measure and monitor compliance.

Contents:

Knowing that the procedures must describe who does what, when and how, they should address these key areas:

Advanced Work Packaging:
- Purpose and Objective
- Definitions
- Contract language
- Creation of the Work Breakdown Structure
- Overview of AWP, IM & WFP
- Key stakeholder deliverables to support AWP
- Project Management Team
- AWP, IM & WFP positions
- Workface Planning software
- 3D model
- Bid assessments
- AWP, IM, WFP kick-off and lessons learned
- Project development roadmap for AWP, IM, WFP
- Optimal path of construction
- EWP, PWP and CWP release plans
- Level 3 project schedule
- E, P & C Work packaging execution strategies
- Audits
- Procedure maintenance

Information Management:
- IM overview
- Execution strategy
- Project nomenclature
- Hardware and infrastructure
- Proof of Concept
- Workface planning software
- Standard model attributes
- Project controls
- Document control
- Procurement and material management
- Model maintenance

Workface Planning:

- WFP overview
- Execution strategy
- Workface Planners
- Installation Work Packages
- IWP release plan (Level 5 schedule)
- Removing Constraints
- Backlog
- Three week look ahead
- Project controls
- Document control
- Material management
- Safety
- Quality
- Field Execution
- Time on Tools
- Level 6 (daily) Planning
- Sub-Contractors
- Testing and Turnover

Contract language: The procedures will become the backbone of the process and will facilitate the detailed description of stakeholder deliverables. It is imperative to include strong language that identifies AWP deliverables in the Requests for Proposals and Contracts. The contract must identify compliance with the AWP. IM and WFP procedures as a requirement. Keeping the details in the procedures and not in the contracts also gives you the flexibility to provide prescriptive methods outside the contract that can be adapted to the current reality as the project develops.

Continuous Improvement: Once your organisation has their first application of AWP safely turned over then the procedures play an important role in the process of continuous improvement by facilitating the lessons learned. Our preference is to call them lessons recorded until the lessons have been incorporated into a procedure, that makes them lessons learned.

Global thought: One of the challenges that we have with the concept of continuous learning is that we often view project imperfections as bad things and they get swept under the carpet for fear of being viewed as personal mistakes. While this 'cover-up' can appear to preserve management reputation, it is detrimental to the process of shared learning through lessons learned. I was told early in my career that I would not live long enough to experience all of the failures that I needed to, so I had better pick up on how to learn from others' failures. I have since found out that this works quite well, but of course only if other people (and I) are willing to share those mistakes.

Flaws in our world of project execution are commonly the result of system breakdowns where well-intentioned project leadership could not see the whole cause and effect cycle, so they continued to do things that were detrimental to the good of the project, thinking that they were doing a good job.

I'll pick on procurement in the following example, but I could have used any department.

The procurement manager knows that he will have to get the module yard to produce and ship 5 modules a month to meet schedule, so he makes this a condition of the contract and uses it to establish progress milestones. The design has modules stacked two high on site with mostly pipe and equipment on the bottom ones and electrical tray on the top ones.

The trouble starts when the mod yard experiences material shortages for the bottom modules. In order to meet the milestone of 5 modules a month they start to develop more of the top modules (easy ones) and less of the bottom modules (hard ones) and they send some modules that are incomplete, to be finished onsite. By doing this they manage to keep the rate of production at 5 and everybody is happy. Except for the people at site who are receiving mostly top modules and some incomplete bottom modules. They have to rent a 100-acre yard beside them to store all of the top modules until they have the bottom modules to put them on. This means that we are now double handling the modules and are tempted to place them out of sequence to take advantage of opportune work fronts, which starts a ripple of 'work arounds' on site. Eroding productivity and adding expensive work hours to the site for the unfinished work.

Now take a step back and ask yourself the question "why do we build modules or create spools in shops"? The simple answer is that it is safer and more efficient / cost effective to build in a controlled environment. So, if:

Yard/Shop = cheap and fast vs

Site = expensive and slow then

What is the reason for creating or tolerating a system that would allow work to be moved from the controlled environment of the yard/shop to the dynamic environment of the site?

If the argument is, that it is because we need a steady flow of work to create continuous work fronts at site, then try this question: Is it better to be slow and expensive than to be cheap and fast, for the benefit of continuous work fronts? Of course not, projects are not developed to keep people employed, (most of the time) they are a means to an end, that produces an operational plant. The problem of inefficient work is created by starting too early and by not maintaining a backlog. As your investment adviser will tell you: The last thing you should do is to throw good money after bad. Fix the problem, don't create new ones.

If this situation sounds familiar, first thing is, don't blame yourself. If you have been involved in one of these rodeos then it was more likely that you were doing what you thought was a good job, contributing to the success of the project. The root cause of the problem is two-fold in that we are hard wired in our roles of project management to get started as soon as possible and that we generally don't understand how systems work.

The object of the project is not to get started, it is to get finished. So now take off your construction hat and put on a shutdown hat. Would you start a shutdown with half of what you need? No way, so don't do it in construction either because you'll get the same result.

Every system is perfectly designed to produce its' result!

This is a profound statement and a bit of a slap in the back of the head because it implies that we go out of our way to design systems that produce the exact opposite of what we want.

We want to reduce the cost and schedule of a project so we design a system that sends modules out of sequence with unfinished scope (that you have to find) to a place where labor is expensive and we have no storage space. WTF? (Water Treatment Facility).

What's missing is global thought. An overall holistic understanding of what the heck we are trying to achieve here (Effective project execution) and then knowing how the parts (E,P & C) contribute to the whole: Systemic thinking.

If this was the result of one of your projects, then this would be a fantastic learning opportunity, if you can find people who are willing to see it as an error and are willing to talk about it. And that's the problem, we don't like to talk about our mistakes or even to recognize that it was a mistake.

When you do manage to develop an environment that lets you learn from your mistakes and fail forward, free of blame, make sure you capture the root cause and then identify the right way to do it in the procedures = then the 'Lesson' becomes Learned.

D. The Work Breakdown Structure

Work Breakdown Structure (WBS) – A hierarchical representation of a complete project with its components being arrayed in ever-increasing detail. The WBS forms a direct alignment between Work, Time and Cost by serving as the basis of Work Packaging, Schedule Development and Cost Coding.

As the name suggests the WBS is a map for the dissection of the project based upon **Work**, not Cost, Schedule, Procurement or Engineering, with the ultimate goal of establishing an alignment between Work, Time and Cost for construction.

It is the very foundation of AWP, Schedule development and Cost management that will either facilitate or derail your ability to understand and manage the project.

The real-life application of the process that aligns work with time and cost happens when we instruct the Foremen to take their IWP number and use it as the cost code on their timesheets. The same IWP number is also used to identify the schedule activity that represents the work. Now you have a schedule activity that is also a package of scope (The IWP) and you have timesheets cost coded against the IWP: Work = Time = Cost.

This gives you a very accurate and simple grass roots process that makes it easy for the Foreman while also producing precise cost and schedule data for project controls. The elements: IWPs, Level 5 activities and cost codes all roll up easily into CWPs to show real time progress and cost performance at a project management level.

The extended application of a good WBS that serves all parties is that it forms the basis for the project nomenclature, which becomes the naming convention for all things. The ideal result is that every schedule activity, IWP number or cost code has this intelligence that corresponds to an area, discipline and window of time. This means that when you are looking at a spool in the laydown yard you could see from its name which CWP and geographic area it belongs to and if you wanted to check it against the schedule you would see approximately when it will be installed.

You could, take a copy of the schedule with you to the laydown yard (which is sorted by CWP) and see which activities have the biggest pile of materials. The same is true for hours burned against schedule activities. Cost code data will show you who is working on what, in the previous week, which when applied to the schedule will show you if we are following the schedule or not and how you are performing against the estimate.

Once you have this working you will wonder why it was so difficult on your last project and why you needed so many project controls people. On top of that, the Foremen will appreciate how simple you have made their lives by getting rid of all those confusing cost codes.

For those of you reading this that have not participated in a mega project yet, I apologize for stating the obvious. But unbelievably this is not how our projects are structured now. Typically schedule activities have a unique naming convention that makes sense only to the scheduler. The activities are not aligned with the way that work is executed. Cost is tracked against cost codes that are developed to satisfy accounting practices and construction do whatever they can and then fudge progress back into cost and schedule buckets.

If it wasn't so sad, it would be funny.

Coming back to the creation of the WBS, there are two critical conversations that you need to have, to get you started on this path of enlightenment, with Engineering and Project Controls. Both departments have a dog in this hunt and both can be very happy with the outcome, but they will need to have influence on the structure. Typically, we start by identifying the way that Engineering like to breakdown their design systems, which may give you something like this:

01- Piling
- 01- Driven / Screwed
- 02- Cutting and Capping

02- Earthworks
- 01- Road works
- 02- Culverts
- 03- Backfill
- 04- Excavation
- 05- Bedding
- 06- Drain Piping
- 07- Liners
- 08- Fencing

Then you can ask the project controls team to identify their requirements for cost reporting. Their standard answer is likely that they need to know how many hours were burned against the elements of the estimate by contract. The extended answer may be that we also want to have a database of installation rates after the project that will help us with future estimating.

Towards the end of the cost code you may find something like this that separates costs by contract and then discipline:

Contract	Discipline	
8114	**001**	**Piling**
8114	**002**	**Earthworks**
8114	**003**	**Foundations**

Sometimes it has another level below this that separates large bore pipe from small bore or heavy steel from light.

For now, let's run with the idea that contract and discipline are the lowest level of dissection.

The levels up stream of this may include cost type: capital or operations and project, cost center etc. All very important to the people managing the purse strings but beyond our line of sight for the scope contained on the project.

The alignment starts when you add construction logic to this matrix and overlay the Engineering and Cost dissections with CWPs (Geographic areas). This is typically when the light comes on for everybody and they realize that engineering, cost management and work execution all have similar segmentation, but slightly different boundaries and names.

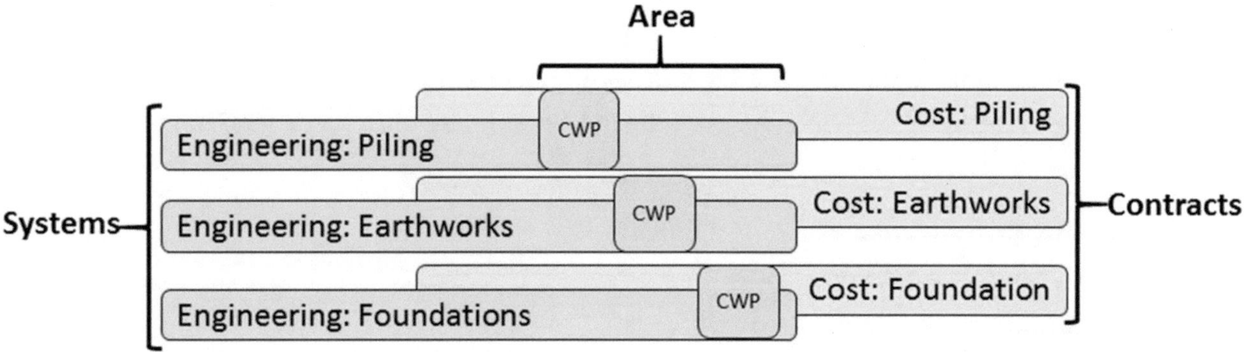

This simple matrix of needs shows that with a little bit of design the WBS can satisfy Engineering, Cost and Work by identifying Construction Work Packages as the common denominator.

To help everybody understand the different levels of the WBS there must also be a WBS library that could look something like this:

Plant: A subsection of the entire project or could be the entire project, a plant is normally identified by the process of operation and geographic region. A combination of facilities that form a unique process. Normally further identified as being Inside Battery Limits (ISBL) or Outside Battery Limits (OSBL)

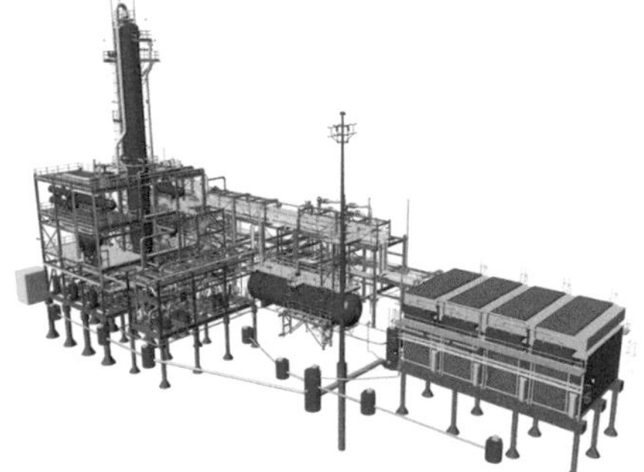

Construction Work Area (CWA) – A geographical division of work defined by Construction. It includes all disciplines, with the exception of cables and undergrounds that are also divided into work areas, but across the entire project. Each CWA has boundaries defined by the logical association of work and becomes one activity on the Level 2 Schedule.

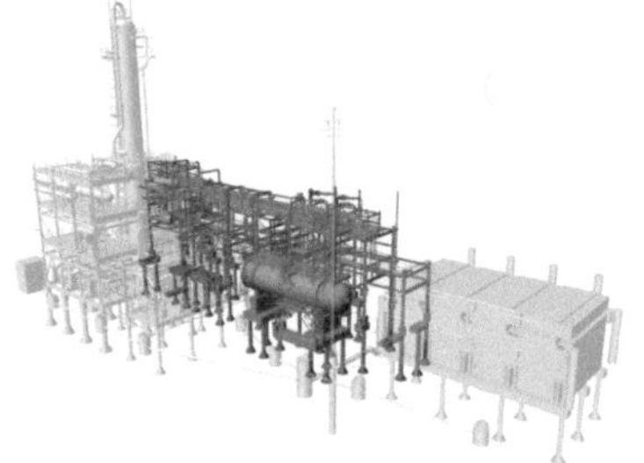

Construction Work Package (CWP) – A single discipline of a CWA that defines a logical division of construction work with less than 40,000 work-hours. A CWP is a component of the WBS, a single level 3 activity on the project schedule and is the downstream product of a single EWP and PWP when prepared for construction. The division of work is defined such that CWPs do not overlap and they can be used as contractual boundaries of work. Each CWP is dissected into a series of IWPs by the Workface Planners.

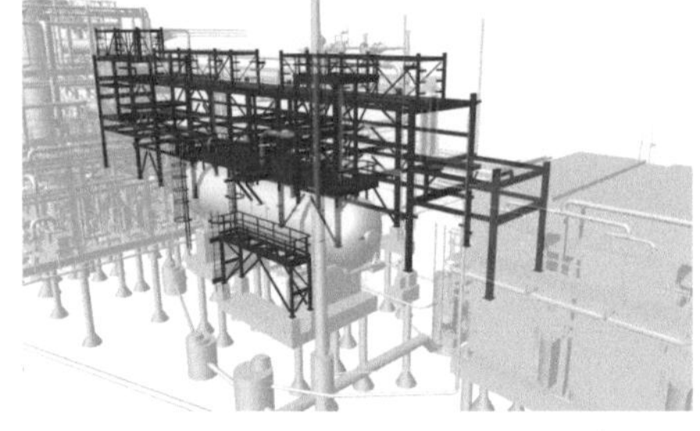

Engineering Work Package (EWP) - An engineering deliverable, single discipline that contains all of the engineering data required for a single Construction Work Package: Scope of Work, Drawings, Vendor Data, Bill of Materials and Specifications, in both PDF and electronic 3D model files. EWPs are developed sequentially to satisfy elements of the Path of Construction, which will facilitate sequential procurement and the execution of CWPs. A single EWP is represented in the schedule as a single level 3 activity.

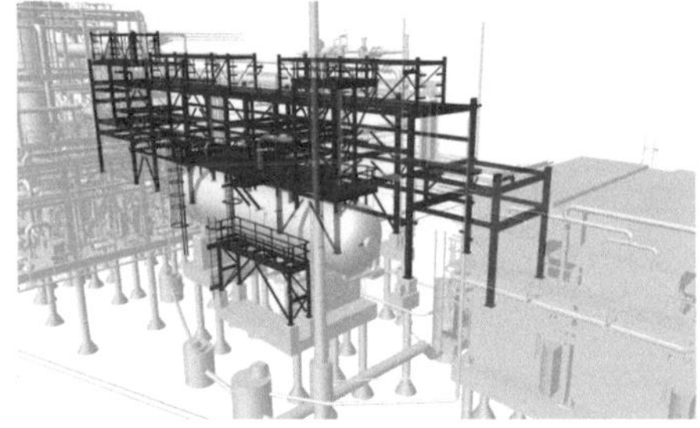

Procurement Work Package (PWP) – A procurement deliverable, that contains all of the materials required to satisfy a single CWP. Typically, a single discipline, in the case of steel and pipe the PWP becomes a discrete fabrication package that is expected to be manufactured and delivered as a distinct group of components.

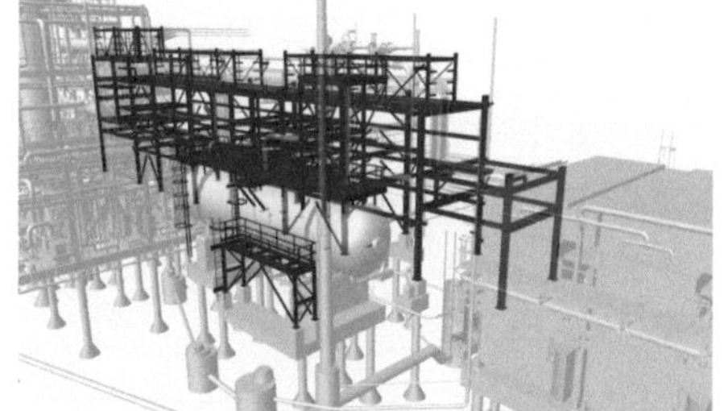

Module Work Package (MWP): A subset of a group of single discipline EWPs that contains all of the Issued For Construction (IFC) engineering data for all disciplines required for the construction of a single module. A group of modules (<10) is a single CWP. The steel and pipe EWPs for a CWP of modules becomes discrete fabrication packages that identify all of the spools and steel piece marks for the CWP (group) of modules.

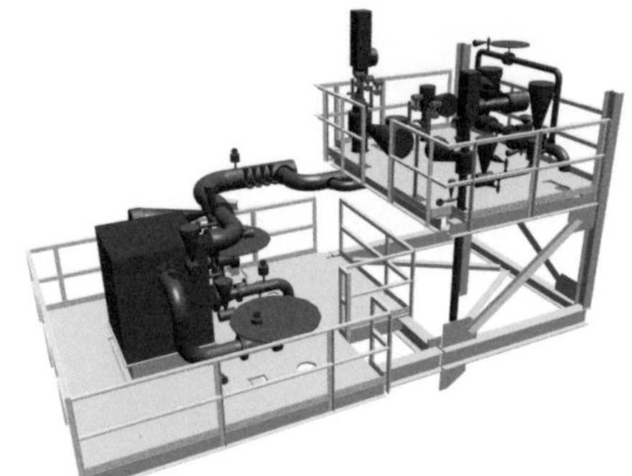

Installation Work Package (IWP): A discrete portion of constraint free, construction work that can be executed by a single foreman and crew, in a single 5 Day period. Dissected from a single CWP and made up of whole drawings. Each IWP becomes a single level 5 schedule activity.

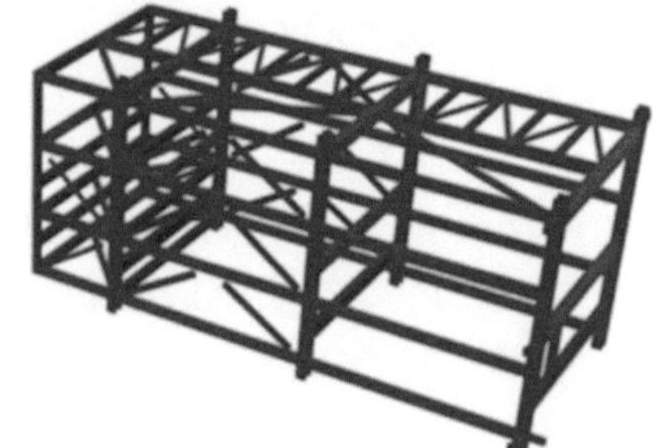

Project Nomenclature:

The creation of the WBS and the development of the WBS library will lead to the identification of the project nomenclature for work packages, which will map the hierarchy of the WBS ancestry. Work packages, drawings, spools, steel piece marks and other components can all use the same nomenclature followed by drawing number in place of the IWP.

WTF-I-12-E4-C05-14

WTF- Water Treatment Facility (Plant)
- **I-ISBL: O-OSBL**
  - **12 - CWA**
    - **E - Major discipline (Earthworks)**
      - **4 - Sub Discipline (Excavation)**
        - **C05 - CWP (C- Construction, E-Engineering, M- Modules, F- Fabrication, P-Procurement)**
          - **14 - IWP**
        - or **12006.1 - Drawing and spool**

Work Breakdown Structure

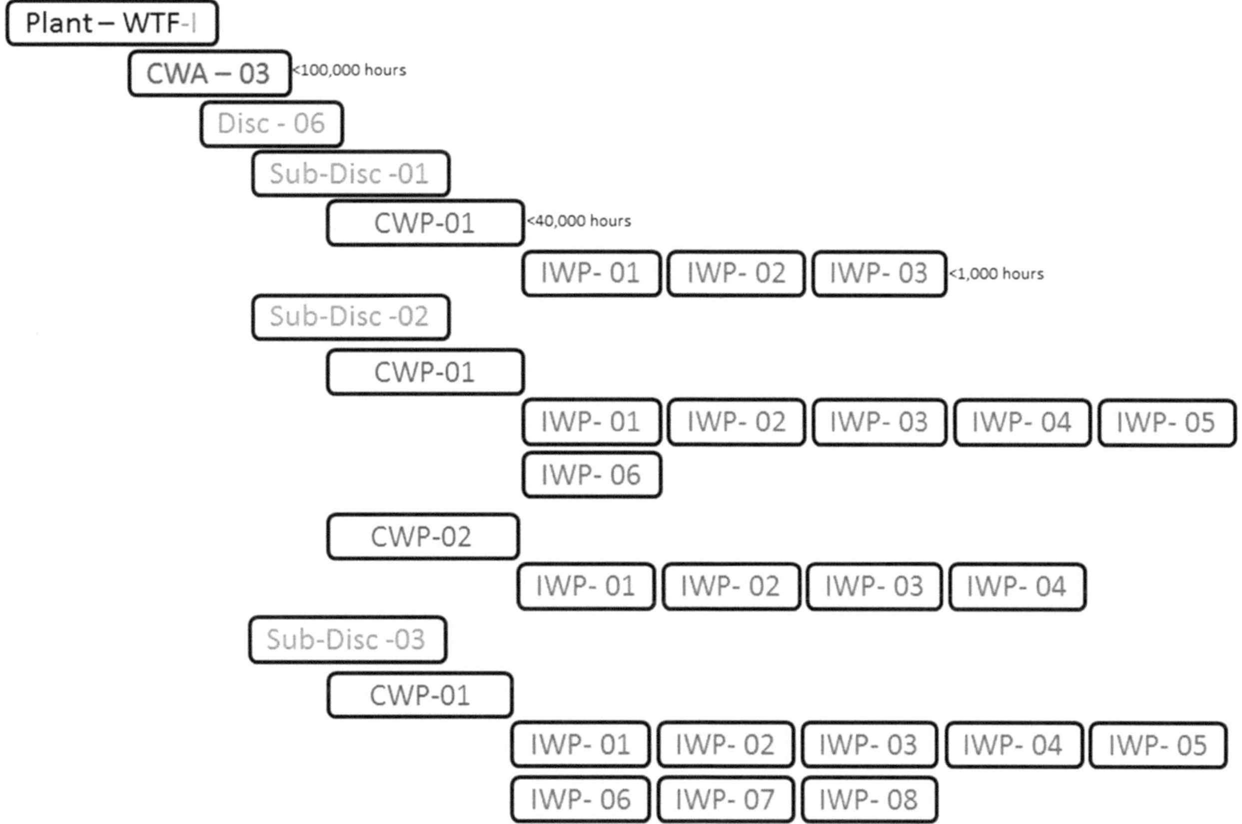

E. Path of Construction (PoC)

The PoC is the articulation of the optimal building sequence based upon the release of CWPs along with the setting of major equipment and modules. It starts during Front-End Engineering and Design (FEED) with the designation of CWAs on the project plot plan and the general flow of work fronts, which typically follows the setting of major equipment and the heavy lift plan. The initial PoC, developed to facilitate the Interactive Planning Sessions, need only contain CWPs for foundations, steel, pipe, major equipment and any long lead items. These are the disciplines that have the longest development cycles (critical path for engineering and procurement). The other disciplines can be scheduled before and after these activities without affecting the start of construction.

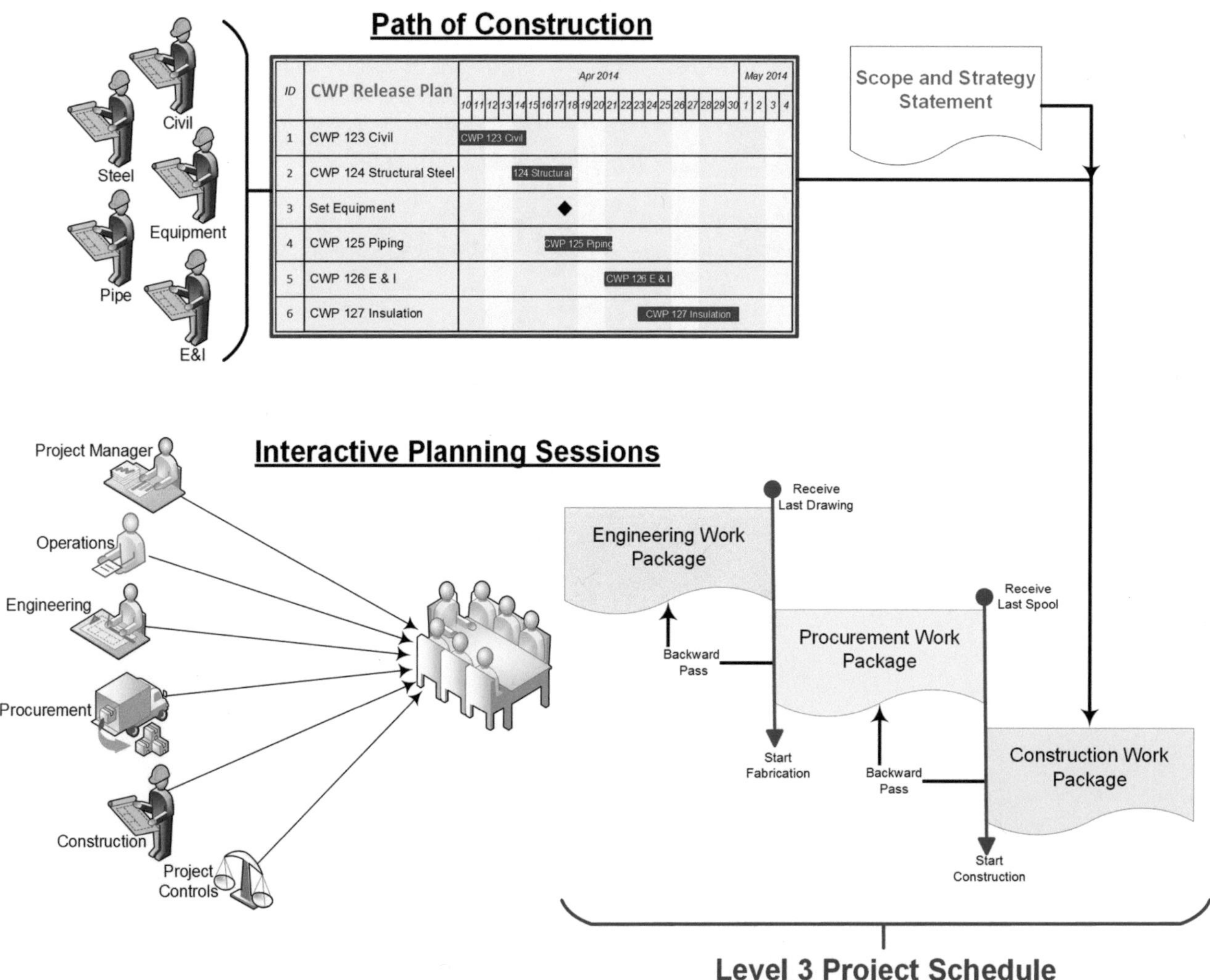

Interactive Planning Sessions (IPS)

The output from the PoC workshop is the CWP Release Plan, which becomes the platform for the IPS. For each CWP start date, the sessions conduct a backward pass that sets a procurement activity (PWP) and duration, which leads to an end date for an engineering activity (EWP). When conducted for each CWP on wall sized schedules with color coded sticky notes, the process brings the project to life and sets target dates for engineering and procurement deliverables. The interactive and iterative process allows engineering and procurement to challenge the sequence and delivery dates of CWPs based upon their own constraints, leading to a final roadmap that links EWPs with PWPs and CWPs in an executable sequence.

This leads us to another significant outcome from the PoC, which is the logic that if we are going to construct by package then we will have to engineer by package (EWP) and procure by package (PWP) and this is typically where we find the first of several rub points.

The nature of engineering follows a systemic development process. You start with a vessel and then run lines to other vessels, this means that you cannot really design by area and then move onto the next area, but that is how it needs to be delivered in an AWP universe.

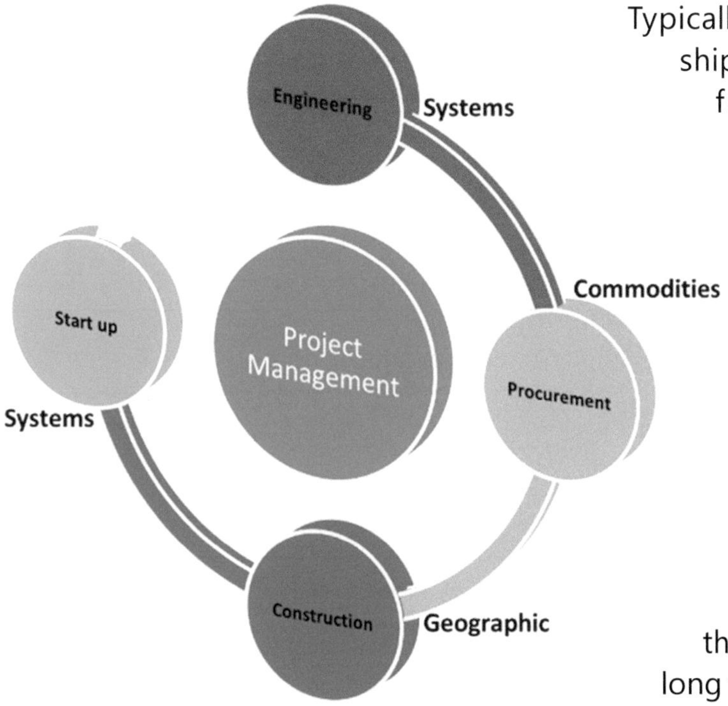

A variation of the same problem is true of procurement. Typically, we procure by commodity, buying shiploads of raw steel and pipe for bulk fabrication. The fabricators then like to fabricate all the heavy steel or large bore pipe first which satisfies production rates but not geographical construction. In an AWP environment, the procurement team can continue to purchase raw materials in bulk as they always have but the final delivery of spools, steel members, and all other components must satisfy the needs of complete CWPs. This means that both Engineering and Procurement face the same dilemma that construction has long faced: Construction like to build in bulks by area, but they have to turnover by system.

Many thanks to my good friend Ted Blackmon for visioneering this dilemma on a napkin for me one night over dinner. He helped me understand that while we do need to optimize production rates for E, P & C by doing things the smart way, we also need to recognize that the deliverables from each of these phases needs to be in a format and sequence that works for the receiver.

Any of these relationships could be characterized as a traditional customer – supplier model, where the customer is the receiver of goods or services from the supplier, which means that Construction is the customer of Engineering and Procurement. In this case, we would identify the Owner as the Client and the project management team would assume the role of customer interface managers.

This is one of the many examples of internal customers and supplier dependencies that we have on projects, where members of the organisation rely upon other people and teams to produce deliverables that allow them to satisfy their own obligations.

In simple terms, we could view Electrical as the customer of Pipe, Pipe as the customer of Steel and Steel as the customer of Civil works. Each discipline needs the previous discipline to do their job well and turnover a product that is fit for purpose so that they can go to work.

This is also true of the relationships that we see in our organisation charts within any single department. The typical craft worker is reliant upon their foreman to supply them with information tools, materials and access so that they can perform their work. The Foreman in turn relies upon their general foreman or superintendent to be their supplier and the Superintendent relies on the rest of the organisation to provide materials, documents, tools, construction equipment and schedules so that they can execute scope.

This logic is the basis of an age-old project management standard that says that if your organisation is not functional, then invert your org chart, (so that everybody can see their customer).

In the same vein, we should view Operations as the customer of Construction, Construction as the customer of Engineering and Procurement and Procurement as the customer of Engineering. This is a good way to describe the relationships that need to be recognized and practiced in a construction driven project.

Many years ago, I was told of a model for commercial construction (Schools) where the construction contractor was engaged as the prime, who then engaged engineering (Design) and procurement. This made the customer also the client, which is a very healthy relationship.

If you look at any highly functional organisation in any industry, suppliers have a very clear understanding of their customers' needs and they shape their products and delivery processes to target customer satisfaction. So now think about that for a moment and try to picture who your customer is and what they really need from you. Now think about what it would be like for you or your department to develop a delivery model based upon this vision of customer satisfaction.

And that's the delivery model that AWP is structured to achieve, with a fanatical focus on construction first and customer satisfaction.

The process of creating an optimal PoC by CWP is the customer telling us what they want and how they want it. Our role as project management, engineering and procurement is to deliver drawings and materials that hit that target.

This is an important point and one that will influence the development of the PoC. It will require good quality facilitation to ensure that Engineering and Procurement understand that their deliverables are changing. This means that at some point they will need to transition from their customary production processes to a new, customer focused, packaged based delivery model.

They may also find that their deliverables are grouped differently. As an example: Construction may ask for the electrical grounding grid as a separate CWP, excluding any other electrical installations, so that it can be installed during undergrounds. Or they may be asked to implement features of scaffold management in the design. These examples are a departure from the existing engineering delivery model, so this important deviation needs to be factored into the contract and highlighted in the Request for Proposals.

F. Level 3 Project Schedule:

The development of the level 3 schedule starts to take shape in the latter half of the IPS. The IPS identifies the delivery date for complete EWPs and PWPs to satisfy the requirement of specific start dates for CWPs.

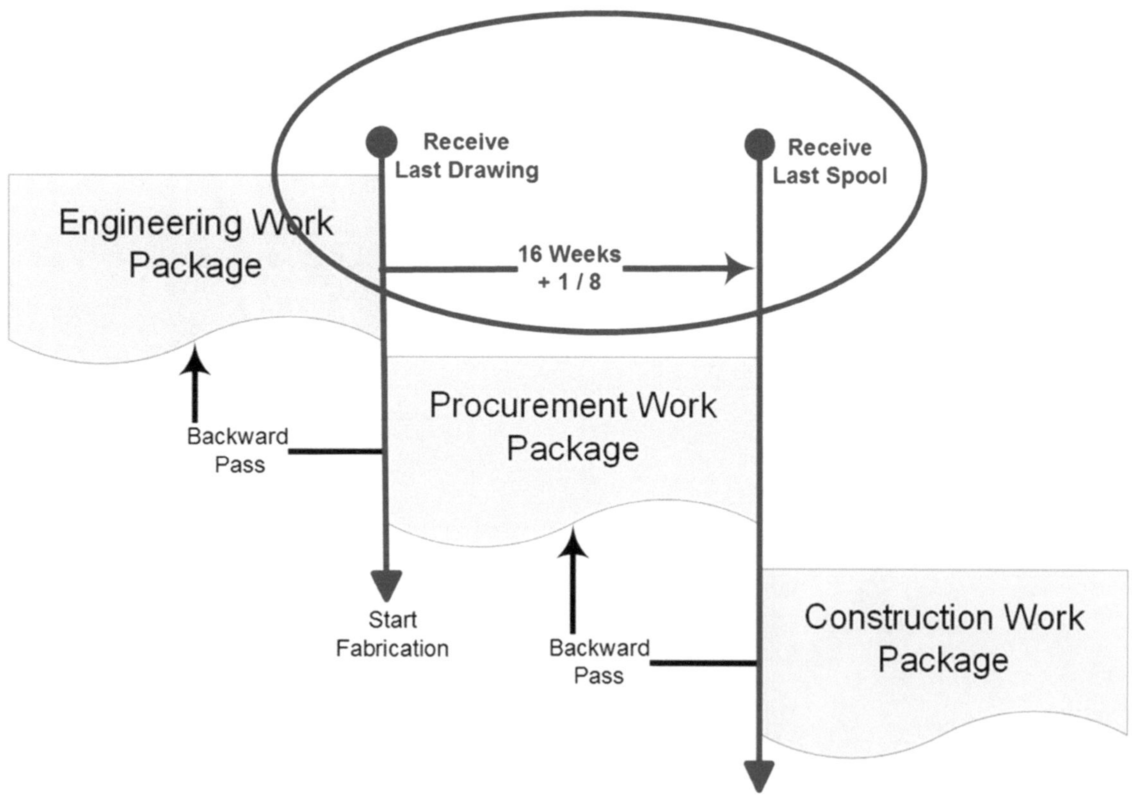

Then with the end date for each EWP (Last Drawing) we work our way backwards to estimate the total duration. The key date in the engineering cycle is the date that the last of the Vendor Data is received. From this point, we can work backwards to identify the issue date of the Purchase Order (PO), which is directly related to the release of the Request For Quotation (RFQ) and the development of Data Sheets by the engineering team.

Moving forward from the receipt of Vendor Data, there is a period of detailed design that leads to a series of 60% Model Reviews, which are the collection point for construction comments and any design refinement needed by operations. After this there is another period of detailed design, that produces the large bore isos and then the 90% Model Review that leads us into small bore isos.

This means that the ideal transition point for engineering to switch from system design to area completion is sometime after the 60% model review.

Knowing that the EWPs will come from model reviews, the design of the model review areas is very important. If we continue to follow the same basic design for plants, of a pipe rack as the spine with plants either side, then the pipe rack is the obvious first choice for a model review area, followed by the plants grouped by design logic and construction sequence.

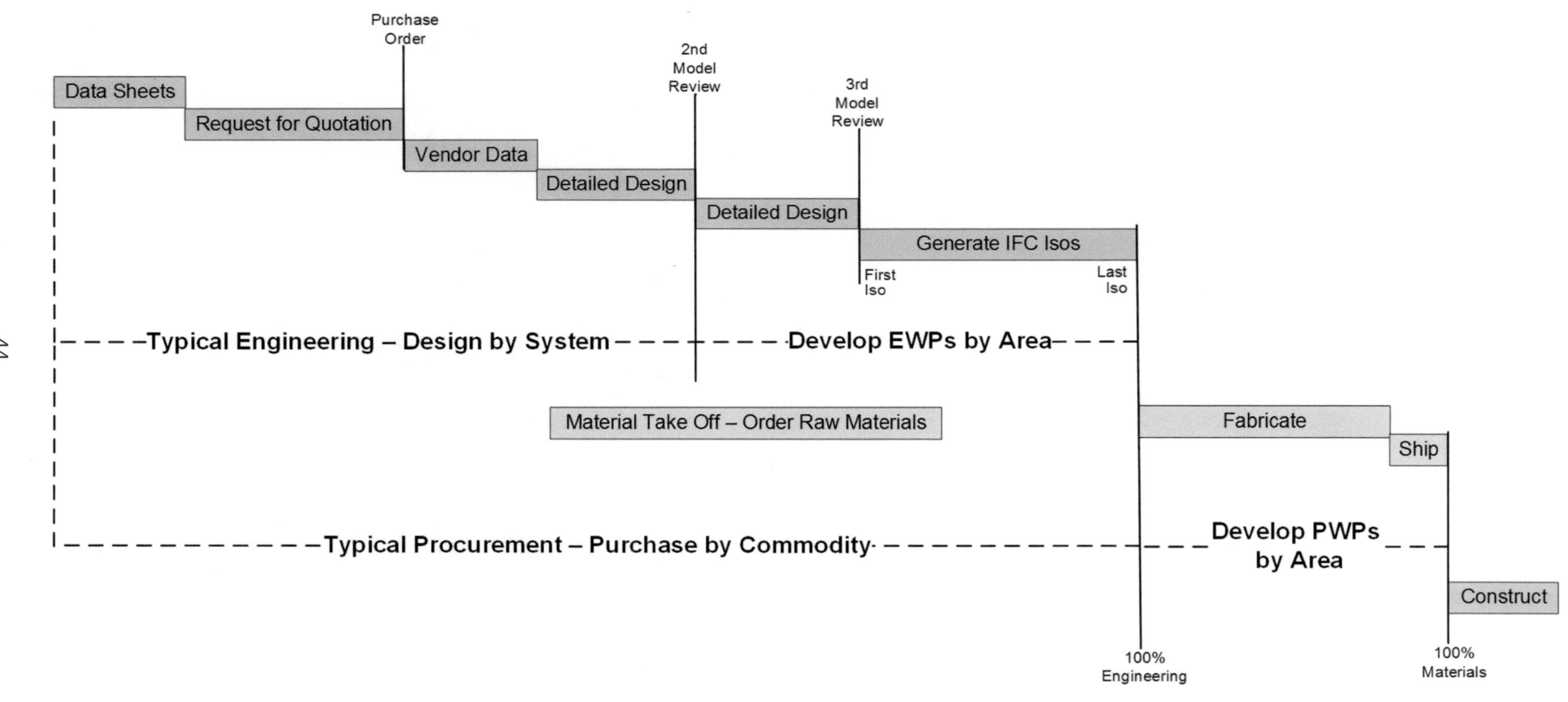
Purchase Order
2nd Model Review
3rd Model Review
Data Sheets
Request for Quotation
Vendor Data
Detailed Design
Detailed Design
Generate IFC Isos
First Iso
Last Iso
Typical Engineering – Design by System
Develop EWPs by Area
Material Take Off – Order Raw Materials
Fabricate
Ship
Typical Procurement – Purchase by Commodity
Develop PWPs by Area
Construct
100% Engineering
100% Materials

The following sample diagram shows that there is a period of development in both Engineering and Procurement where the organizations operate based upon their preferred (existing) processes: design by systems and purchase by commodities, then they switch to a focus by area and deliver single EWPs and PWPs that satisfy CWPs. All of these factors need to be considered by Engineering and Procurement as they are developing these durations and delivery dates.
The anchor point for the development of all these dates is the start date of the CWP. From this point procurement will determine a cycle time for the fabrication and delivery of Steel and Pipe. A backward pass then gives you a start date for fabrication which is the finish date for engineering.

Important note: The relationship between EWPs, PWPs and CWPs is strictly Finish to Start in an AWP environment. That means that you cannot start fabrication until the last drawing for an EWP has been released and that this has triggered the release of the complete EWP. No more bootlegs or building from IFRs (Drawings Issued For Review).

This is also true for procurement, the delivery date that they commit to, represents the delivery of the last spool or steel member, which triggers the start of construction.

The most effective way to drive these milestones is to track progress towards them (in the WFP software) and then link the achievement of milestones to payment points in the contract.

So where did the start date for each CWP come from?

There are a couple of ways to develop a level 3 EPC schedule: You can 'Start with the end in mind' by having a delivery date for the project and then working your way backwards. This will show you if you are starting this exercise in time or if you have missed that train. Or you can start with a mythical date (Jan 1st2000) and develop a project schedule that is not anchored to a real start date. This would give you a window of time, like a shutdown, where the activities are all linked to each other to represent the total duration of the project. Of course, this only works if you are using scheduling software that will link all of the dependencies and allow you to insert a real start date later, which bumps all of the activities but keeps the right duration. Primavera or Microsoft Project can both do this.

Either way the construction team need to do the hard work of developing their optimal PoC by having their Subject Matter Experts (SMEs) develop analogous estimates for CWPs which are then sequenced based upon leads and lags. This is a good place to use quality facilitation to make sure that the CWP release plan actually represents the best-known combination of events. The SMEs don't always know how their activities relate to other disciplines.

Let's skip forward to the end of the IPS, you now have a roadmap that links each CWP to a predecessor PWP which has a predecessor EWP, with a WAG estimate for durations. These elements are the ideal components of the Level 3 schedule, don't be tempted to go into any further detail, this is all that the project schedule needs. Internally E, P & C can develop their own details as they see fit, but at a project level, this is as far as you should go.

Looking at the project holistically, the layout of scope as dependent, finish to start activity streams in this format: EWP-PWP-CWP allows the project to consider the parallel execution of multiple streams, which is the footprint for Fast Track construction (without the chaos).

It is now the responsibility of the E, P & C teams to flush out their work packages and refine their duration estimates based upon quantities, as they become known.

This also means that the project schedule is a living document that is updated each week with new EWP, PWP and CWP durations, with a target for the level of confidence that is required by the end of FEED (+/-10%?).

We will spend some more time on WFP software in the Information Management section, for now it is enough to know that this software applied during FEED is a simple and fast way to extract weekly CWP quantities from the model (which feeds the E, P & C estimates).

G. Work Packages for Engineering, Procurement, Modules and Construction.

Construction Work Package: On a typical project, there are several key interface points between Engineering, Procurement and Construction. The principal one for AWP and WFP is the Construction Work Package. As discussed earlier, it is the common denominator for E, P, C and project controls. It becomes the target for Engineering and Procurement deliverables, the starting point for Construction, the job description for the Construction Management Team and a very effective way to issue work to the Construction Team or Sub-Contractors. It also works very well as the anchor point for project controls in both schedule and cost management.

For our purpose, think of each CWP as a mini project. It is the center of the project universe and the critical link between all the upstream processes and downstream construction. It is that mid-point in each mini project cycle.

We have seen examples of projects that have advocated that the CWP is not necessary and that the EWP can be issued directly to construction. In the short term, this looks like it works but the reality is that it just moves work and responsibility away from the CMT and onto the desk of the Construction Contractor. This increases the work load of the Contractor, who have less access to the data or authority that is needed to develop the rest of the CWP, which ultimately detracts from our end goal of facilitating the foreman.

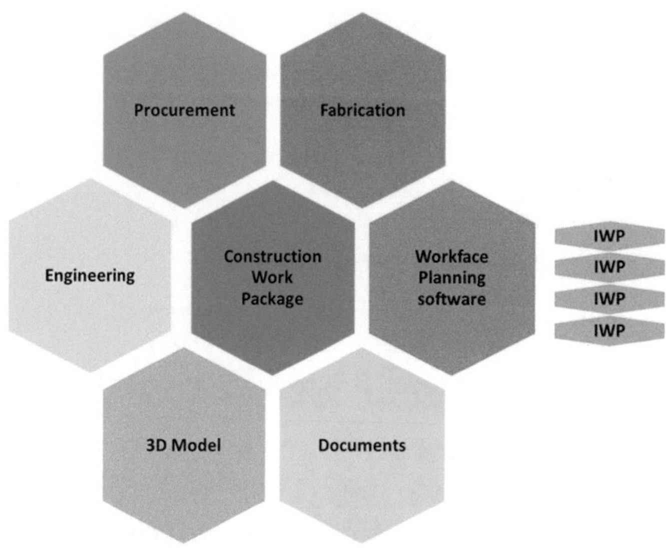

Contents of a CWP:

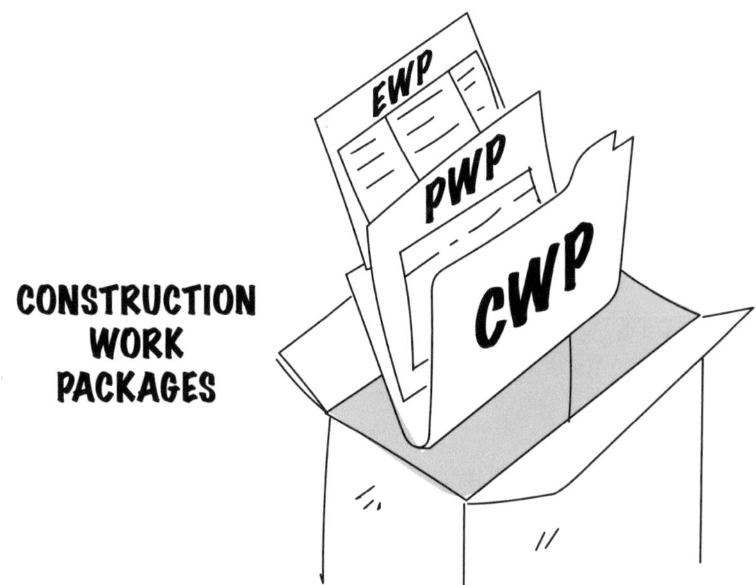

As an instruction of work the CWP must contain a complete description of the scope, a detailed list of the support services that will be required and a definitive estimate (+/-10%) of work hours and duration.

A typical CWP of 10,000 - 40,000 hours:

Scope: A detailed narrative of the scope with engineering drawings, 3D model shots and plot plans. This section should also identify work that is excluded from the scope and must contain an execution strategy that identifies how the execution of scope is envisioned with interfaces to other disciplines.

Rigging/Lift studies: If the scope identifies critical lifts that require a lift plan or rigging study then the CMT are responsible to conduct the study and insert the final report with instruction into the CWP. Rules and standards for rigging can also be inserted here.

Project Controls: Copy of the level 3 schedule highlighting the CWP's and the relationships with other CWPs or activities with estimated work hours and a duration. The combination of work hours and duration should also be displayed as a Craft Resource Plan.

Safety: A high level Safety Plan that identifies the known hazards and mitigation strategies can be developed early based on estimated work, location and resources. This can include links to a library of Safe Work Practices that the CMT have developed and presented to the Contractor. The plan identifies the training for craft, classifies the permit requirements and describes the process for obtaining permits.

Quality: Can be a specific Inspection and Test Plan (ITP) that covers the unique scope of the CWP or could be a description of how the scope is covered by standard ITPs that are available through a library supplied to the contractor by the CMT.

Construction Equipment: A detailed estimate of the specific construction equipment that will be required to execute the scope. Will include cranes, elevated work platforms, welders, pumps, generators and special equipment not supplied by the tool crib. With a description of whether the contractor is expected to supply their own equipment or if it will be supplied. Must contain a description of how to order and secure supplied equipment.

Scaffold: A detailed estimate of the scaffold requirements needed to execute the scope in a 3D model shot along with the overall project scaffold strategy. Instructions on how to order scaffold and an estimate for the building cycle that is required.

Material Matrix: Extracted from the 3D model the material matrix must list every tagged item including spools and steel members, with a list of bulks and all other materials. Each component is identified as being supplied by the project or by the contractor. This section will also contain a description of the material receiving process and instructions on how to determine if materials have been received and inspected along with the process for requesting materials.

Squad Check: This can be viewed as a group activity that is the initial step in the process used to deliver the CWP to the contractor. Typically, the CWP is sent to the contractor as the scheduled execution date enters the 90 Day window, with an invitation for the Squad Check to take place within one week. The squad check team includes the Superintendent, their Workface Planner, the CWP Coordinator with representatives from engineering, construction management and project controls. The 1-hour process walks through each section of the CWP and records questions and comments.

The Superintendent and the Workface Planner then work together to decompose the CWP into Installation Work Packages based upon the Superintendent's execution strategy.

Typically, templates are developed that cover the generic information for CWPs

Engineering Work Package: Section E. has a simple flow chart that shows how the process of work packaging starts with the creation of the Path of Construction. This is best completed with support from Engineering, who run the model and divide the plot plan into CWAs based upon the logical association of work identified by Construction. As each CWA is then split by discipline, it creates individual CWPs and this forms the geographic definition of Engineering Work Packages that match the CWPs one to one.

Contents of an Engineering Work Package:

Scope: A detailed narrative of the scope with an implementation strategy that identifies how the execution of scope is envisioned with interfaces to other disciplines. Should identify the work that is excluded from this package.

Engineering drawings: A complete list of the IFC drawings and plot plans for the entire scope, installation manuals for vendor equipment and directions on where to electronically access the drawings. (Document Control database on the project cloud)

3D model: If not already available through the cloud, then the fully populated 3D model for this EWP is to be made available on the cloud. The hard copy EWP should contain model shots that give a high-level overview of the scope.

Level 3 project schedule: Isolated extract from the Level 3 schedule that shows this EWP with links to the specific PWP, the CWP and any other dependent EWPs.

Material Matrix: A detailed matrix that shows a list of all materials required to construct the CWP, the organization responsible to procure the components, the Fabricator/Supplier and RAS date (CWP Start date), if they are known.

Holds: While the criteria for delivery of an EWP is that 100% of the drawings must be available, there will be cases where this is not practical. In these cases, the EWP should contain a detailed list of holds, with the person responsible for managing the issue and a forecast for resolution.

Procurement Work Package: The identification of a package of work for procurement is a developing conversation within the industry. We know that there is a gap between the end of the EWP (last drawing) and the start of the CWP, which is where the procurement process takes place, but the idea of a unique package for this procurement activity is not common.

The place to start this dialog is with the logic that we know we need all of the material onsite before we can start any single CWP (last spool), so the question is where does this list exist and who is managing the procurement of the components against this requirement?

The ideal answer is that the Procurement team have identified all of the materials required for a specific CWP as a unique package of work for themselves, which allows them to manage fabrication and material purchases against this Required at Site date (which is the start date of the CWP).

The way that this has manifested for us is that the Engineering team produce an EWP and then extract the material matrix and fabrication drawings to create a PWP.

The procurement team normally have a head start on this list, having generated a material take off from the entire model sometime before this point, which led to an order for ship loads of raw materials for steel and pipe and any long lead items.

The list of components for a single PWP is entered into the procurement software against Purchase Orders with a common PWP number and a common RAS date. This allows the procurement team to manage fabrication and bulk procurements at a rate that satisfies the PWPs.

For bulk purchases (piles, valves, cable etc.) the soft allocation function in the procurement software shows whether there is enough stock onsite to satisfy the PWP RAS date.

The process of fabricating all of the steel or pipe for a single CWP is the primary target of PWPs. While the process is sub-optimal for fabrication, the delivery of complete PWPs facilitates productive construction. This means that the fabrication package that goes out must have strict guidelines that ensure that the PWP is delivered complete, by the RAS date.

Contents of a Procurement Work Package:

Scope statement: A description of the procurement scope and strategy for the acquisition of components and the fabrication of equipment, steel or pipe, along with the common RAS date. This is also a good place to reiterate the requirement for electronic fabrication data (spool numbers and piece mark numbers), the tagging of components with barcodes or RFID tags and the reporting of weekly progress.

Material matrix: An extract from the EWP the material matrix shows a list of all materials required to construct the CWP, the organization responsible to procure the components, the Fabricator/ Supplier and RAS date (CWP Start date).

Fabrication drawings: A component of the EWP that is the complete list of equipment, steel or pipe drawings that need to be fabricated, for a single CWP.

3D model: The section of the 3D model that is made available on the cloud, through the release of the EWP, should be offered to the fabrication shops so that they have the option to import the data into their own software.

Module Work Package: There are two types of work packages that influence module construction:

- The individual module, which contains all of the requirements for a single module.
- A group of modules that form work fronts onsite, which is a CWP.

While it's a common practice to identify each module as a CWP, this does not satisfy the criteria for CWPs in the world of AWP. For Construction, a single CWP is an association of scope that opens a complete set of work fronts in a single area. If you talk to your construction folks and ask them which groups of modules they would like to receive so that they could set them and open work fronts, they will divide the modules into groups of 10 or less in areas that supports the construction strategy.

The logic of a group of modules forming a CWP also works well for the fabricator and the Mod Yard. Rather than fabricating the pipe for a single module, the fab shop can fabricate the pipe or steel for a group of 10 modules, which gives them some level of optimization. The same is true for the Mod yard, knowing that these ten modules have to be completed at the same time the yard management can share crews across all ten modules, which supports the logic of layered construction, which is how modules are built.

Having said that, we still do need to produce a single package of multi-disciplined work for each module: Module Work Package. The good mod yards park a sea-can beside the module and when the complete list of materials has been received then the work can commence.

Contents of a Module Work Package:

Scope: A detailed narrative of the scope for each discipline with engineering drawings and 3D model shots, separated by layers. This section should also identify work that is excluded from the scope (ship loose) and any shipping steel. Must contain an execution strategy that identifies how the execution of scope is envisioned along with the requirement for Workface Planning and Installation Work Packages managed in WFP software. Identifies all of the modules in the same CWP group.

Project Controls: Copy of the level 3 schedule highlighting the Module as a component of the CWP and the relationships with other CWPs or activities with estimated work hours and a duration.

Quality: Can be a specific Inspection and Test Plan (ITP) that covers the unique scope of the module or could be a description of how the scope is covered by standard ITPs that are supplied to the module yard by the project.

Scaffold: A detailed description of the permanent scaffold requirements (scaffold that will be shipped as part of the module) with 3D model shots along with the overall project scaffold strategy.

Material Matrix: Extracted from the 3D model the material matrix must list every tagged item including spools and steel members, with a list of bulks and all other materials. Each component is identified as being supplied by the project or by the contractor.

Driving Packaged Behavior

The philosophy of delivering work packages rather than single Isos, spools or modules and the process of executing single CWPs as unique projects, is the foundation of AWP. This is also the single biggest change that your projects will face. The process sets up a flow and a set of rules that people will gravitate towards when they understand that it makes sense. We all like a little bit of structure and a clear understanding of what to expect and to know what is expected of us. The dependency relationship between E, P & C work packages has a very simple alignment and a simple set of rules. Finish engineering and start procurement, finish procurement and start construction. However, getting to this point does take some hard work and discipline.

In any change process, we travel through a series of questions: Why, What, Who and eventually How. When you get to How, one of the ways to drive this behavior is contractually by establishing rules of credit based upon delivery of complete packages. In our current execution model, we set up contracts that establish targets based upon the number of drawings delivered IFC or by tons of steel or meters of pipe delivered per month. If you want to drive the philosophy of packaged deliverables then you need to establish credit milestones based upon completed packages. This means that Engineering and Procurement only get credit or payment for the last drawing in each EWP or spool/steel member in each PWP.

This is also true for modules. Construction don't need a set number of modules per month to stay busy, they need specific groups of 5 or 10 modules that open work fronts in specific areas, so that is how the contracts need to be structured. Then milestone payments should be based upon delivery of the last module in each group.

If you set up a contract for a fabricator where the delivery of all of the steel members for a specific CWP triggers a payment then that fabricator will find the best way to achieve that target. Be warned that this will also mean that their suppliers will be under the same pressures to deliver bulk material in the right sequence. If the bulks are being supplied by the Owner then they will have to be on their A game to ensure that they are not getting back charged for causing delays. (Which is also a healthy relationship).

The same can be said for Engineering, an Engineering firm that gets paid to deliver whole EWPs rather than individual drawings, will be driven to reach that milestone by designing by system and then switching to area completion at the right time.

You are probably thinking right now that this is going to cost you, and you're right. The Engineering and Procurement companies are likely to ask for higher rates to cover the unproductive behavior that comes with designing by area or retooling the steel mill every week. However, the rate of return for this investment is very attractive, my rule of thumb, based on non-scientific WAG data is that 1 hour spent on engineering or procurement saves 10 in construction. (The construction hours also cost a lot more).

Bringing this back to the conversations that need to take place during the PoC, you will need to establish the delivery model for the Engineering and Procurement teams early so that they can get their heads around what they will have to do to get paid. This will probably also drive the idea that it is better for packages to be smaller than bigger. The ideal sweet spot for Construction Work Packages is in the range of 10,000 to 40,000 hours in a single discipline.

The only other consideration that you should address is that there is no such thing as 100%, so when you set that as a milestone you will also need to develop some criteria that allows a bit of wiggle room. It could be something like: Less than 2% of drawings with holds and an error rate of less than 2% on spool dimensions. With exceptions for vendor data or long lead materials.

If you don't have this exception clause then the project could be log jammed waiting on a couple of 1" drain valves.

Perfect engineering or procurement is far too slow and expensive to support the principles of AWP. The strategy is to get as close to perfect as possible without paying too high a price through schedule loss.

In summary, an alignment of Engineering with Procurement and Construction that targets Customer Satisfaction, is the model that is missing from our projects right now. The silo mentality where each department is responsible to optimize their own system without much regard for how their deliverables contribute to the whole project, is an antiquated thought process that has escalated costs and elongated schedules on many projects.

An important note here is that these changes must be driven by the Owners. If we take a look back to the dramatic changes that we experienced with safety performance over the last 30 years, it is easy to identify that it was the Owners' influence that drove behavioral changes. The same is true of productivity and maybe even more so. The losses experienced in a productivity failure rarely effect the Contractors.

When it all comes down to it, the applied function of what we do on projects is that we give the foremen information tools, materials and access, so that they can manage the process of construction for us. The model for AWP defines a line of sight between the generation of information and materials against that end game.

I heard a story once that the cleaning lady at NASA in Houston was asked what she did for a living and she replied that she was part of a team that put people on the Moon. That is what AWP is searching for, that day when a junior engineer in a design team in India understands how their contribution facilitates activities at the workface.

Chapter 6: Information Management

It is not always easy to see ourselves as being the pivotal players in a revolution, but that is where we are. We have the history of cavalier construction behind us and the age of agile, lean, professionally planned, predictable project execution ahead of us. The spark for this revolution is the need for change, the gasoline on this smoldering kindling is free flowing information and the technology to do something with it.

Welcome to the golden age of information management for construction.

Most of our construction people live Jekyll and Hyde lives right now, where their 'normal' persona is the typical construction worker, who stands around waiting for scraps of paper that contain information that allow them to get small amounts of work done. Then in the evening their ultra-ego is released and they become information junkies, communicating with the world, surfing the web, consuming news at the speed of internet (much faster than the speed of light).

I have often been frustrated when caught in this cycle, where you know that the information you need for construction is out there somewhere, but you cannot access it because the handlers don't know how to post it, or even that somebody else would like to get it.

At any given time on our projects somebody knows the current state of where every piece of material is, how to get the latest revision of each document, what has been installed, what we should be installing, which resources are available, what our internal customers need from us and how much time and effort we are using to get stuff done, but we don't have this information in a managed environment. It's like we are each writing chapters of a mystery novel that we are not sharing with each other, the parts make sense individually but nobody knows the whole story.

It is this labyrinth of loose ends that leads us to a state that begs for organization and ultimately to a diagram that connects the generation of information with the end users.

In the first book, Schedule for Sale, we talked about the transition of data to information to knowledge and finally into a state of understanding. Nothing has really changed; the steps of transition are still the same that they were 10 years ago, however, time and firsthand experience have given us a better appreciation of the detail and structure that is needed and that has led us to this road map.

INFORMATION MANAGEMENT

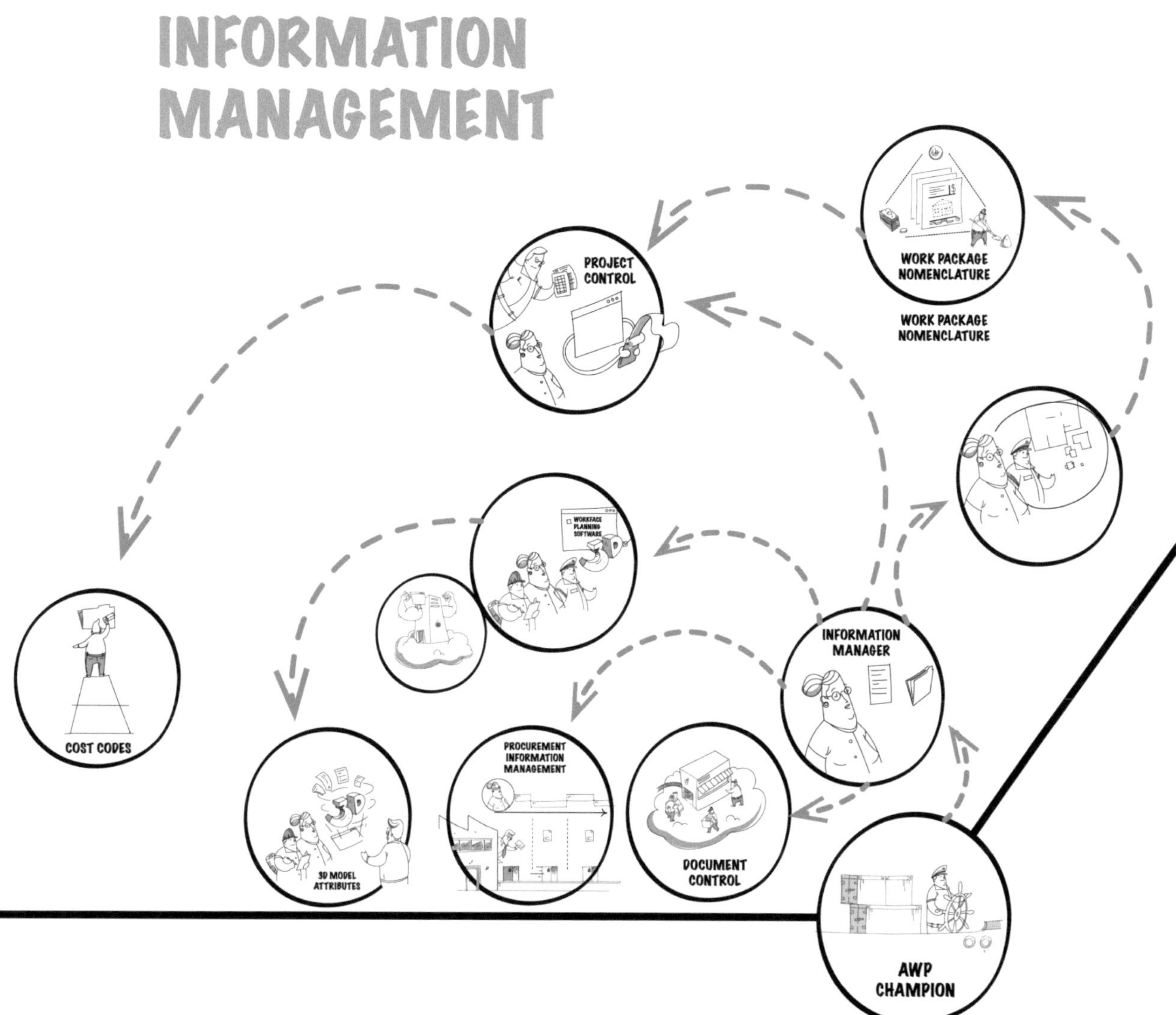

This chapter will lead us through the details of each node on the map and hopefully make that connection for you that links the generation of information with the method of delivery and the way that the end user will access and apply it.

A. Information Manager
B. Work Breakdown Structure
C. Workface Planning Software
D. 3D Model Attributes
E. Document Control
F. Procurement Information
G. Work Package Nomenclature
H. Cost Codes

A. Information Manager:

The first thing to think about is the separation of Information Technology (IT) and Information Management (IM). We commonly think of IT as the computer guys who create our user names and passwords, manage hardware issues, install software and magically fix our computer timeout issues.

IM on the other hand, is the orchestrated development of project information so that it is fit for purpose and beneficial to the end users.

So, from this point forward I want you to be suspicious anytime that you hear that there is an IT/IM manager on your project, they are totally unique roles and it would be a rare person that could understand both fields.

In our infographic, we identify the Information Manager as a position that reports directly to the AWP Champion, which is very important and telling of their role. A high-level description would be that the position is responsible to develop the processes that will make the WFP software fully functional. When you take a hard look at what that will take, you will find a definitive set of standards with complex interface points, that must be spelled out in contracts, taught to the information generators and continuously monitored and re-aligned over the life of the project.

This is probably one of the hardest positions to fill, you need somebody that has an acute understanding of WFP software, 3D modeling and AWP, with a general comprehension of engineering processes, supply chain management and how to operate in a cloud environment.

You can find this expertise in different pockets, the WFP software vendors can supply expertise on their software and the engineering companies typically have in-house expertise for 3D modelling, but you still need somebody who understands the big picture and how to bring the parts together. So, as I said in the first book, if and when you do find this person, treat them well and offer them a company tattoo.

In our most recent full-blown applications of AWP we had a database guru riding shotgun with the IM. This creates a two-pronged offense that can work together to create the harmony of the right information in the right format.

Information Manager: Job Description

The idea that the IM is responsible to make the WFP software work is a very simple way to explain the complete process. The complexity starts when we get into the weeds around, what does it take to make the software work? The functionality of the software is wholly dependent upon the quality of data. Typical inputs for the software are all of the other headings in this section, along with some other 'must haves' that we will explore.

Let's start with the Cloud:

We don't talk much about this in the infographic because it is infrastructure rather than process, but it is a critical strategy that can make or break the whole program.

Think back to your last project where the Engineering, Procurement and Construction Management teams were a separate company from your constructors. There was probably a gap in the chain of information handover where data was delivered as hard copy PDFs.

Typically, this is engineering drawings emailed as non-intelligent PDFs, packing slips with piece marks or spool numbers, hard copy vendor manuals or hard copies of 3D snap shots. While this has the appearance of satisfying the requirement of a deliverable, most often the end user has had to recreate the data or engage resources to manage the information in a field office.

The problem of 'almost good enough information' stems from an old misconception that construction companies lack the sophistication or desire for live information, which is a good example of a self-fulfilling prophecy. Our organizations thought that the constructors didn't need or want live data, so they didn't even offer it. The Contractors adapted to a world of hard copy information, strapped to a piece of pipe and made up systems to manage whatever data that they could get their hands on... and the prophecy is proven.

Now the world outside of the construction community (the internet) has shown contractors that there is an abundance of information that can be readily available in a format that is aimed directly at satisfying customer's needs. This has created a very low tolerance amongst constructors for the existing system of mushroom management. (kept in the dark and fed on manure). The awareness of knowing that other options exist take people from being ignorantly content to a world where 'wants' have become 'needs'. This welcome change is driving the recognition and respect that we should have always had for the end users. Evolution has demonstrated that humans adapt well to changing environments and we have definitely seen construction organizations step up when we give them direct access to live information.

The existence of drip fed, hard copy data in the past has also prevented us from understanding the need for effective delivery models and a common project platform that could be used for the electronic exchange and storage of project data.

Typically, we look at the Engineering and Owner's networks as the right platforms to house the systems that we need to manage the project with the thought that they are the only people who need to access the information. Then we get to site and realize that we need to give and get information to the contractor. You can build work arounds with Share Point, Dropbox or FTP sites but they are not really fit for purpose and you will spend lot of effort and energy creating workarounds that get worked around. Eventually the project exchanges information by email or hard copy that gets stored in countless spreadsheets. Apart from the significant loss of time it takes to do this we also create a series of alternative facts, which is its' own problem.

The answer that has worked for us in the past is to think holistically by creating a cloud based project environment that is the single platform for all the project data interface points. This allows the project to create a single copy of the 3D model, documents and procurement data which is then made selectively available to the project users through a secured environment.

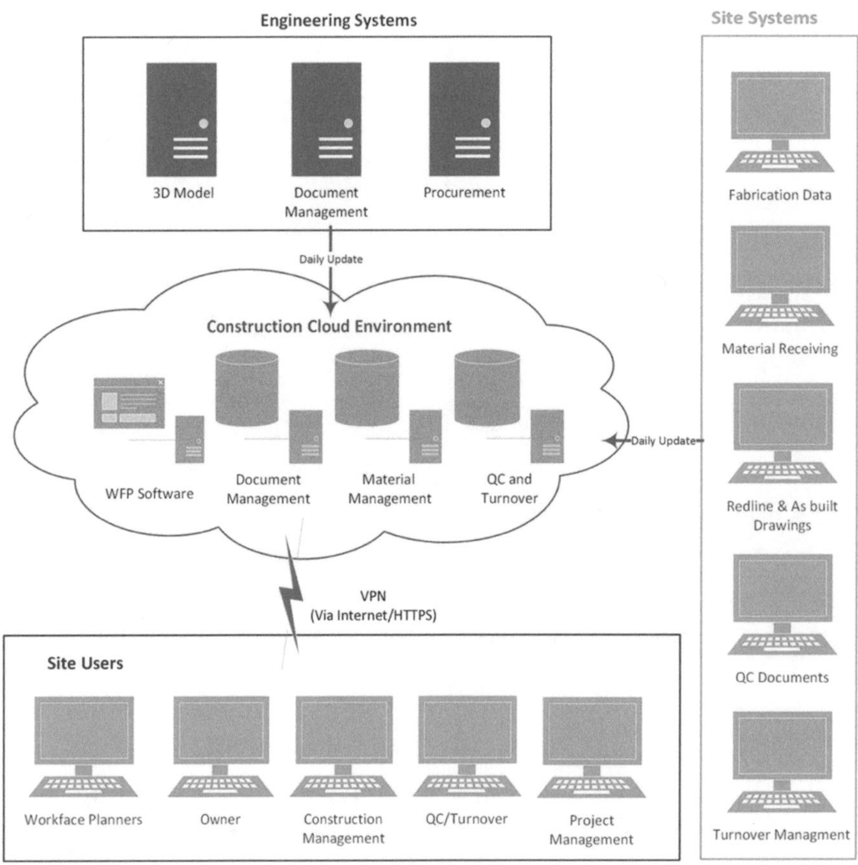

If you have access to the internet and a user logon, then you have access to the project information that you can access from anywhere. This keeps the mothership networks isolated from the project, while still facilitating the delivery and exchange of live project data.

This transparent data centric model for project information also supports the idea that the project management team should manage the project by having direct access to the project data.

In our typical current project environments, the Project Management team mandate standard reports that tell them the state of the project. The reports are generated by the many stakeholders drawn from their own inhouse data management systems. Then 'management of the message' creates a system where everybody develops their own report card.

Is it any wonder that projects are full of surprises?

In a data centric model where the project management team manage the master interface in a cloud environment, each stakeholder reports their granular information into the cloud and the Project Management team design and draw their own reports. On the projects where we have set this up the Project Management team would develop reports and send them to the contractors, showing the contractors their progress. The end result was that we had very few surprises and we learned to appreciate that 'bad news early is good news'.

B. Work Breakdown Structure

WTF-I-12-E4-C05-14

The WBS chapter in the AWP section of the book gives a comprehensive review of what, why and how the WBS is created, suffice to mention that it is the backbone of the WFP software and if you don't get it right, don't bother going any further.

Importantly the design and administration of the WBS is one of the key functions for the Information Manager. Working with all of the stakeholders they are responsible to establish a single format for all project users so that the CWP number appears in the name of every work package, drawing, spool, steel member, schedule activity, and cost code.

C. Workface Planning Software

Our criteria for WFP software is that it must:

- Integrate with a **3D model**: Operate in a 3D environment
- Allow users to **build IWPs** in a 3D environment: click on objects and add the data to a plan.
- Be able to develop **tasks** for IWPs: Choose an object and select a single task for the component (weld)
- Can accept **fabrication data**: Permanently add spool and steel member numbers to objects.
- Be able to calculate **planned value** for tasks or groups of tasks based upon administrator defined installation rates and rules of progress: Tell the planner how many hours will be earned when the task is complete.

- **Link drawings** to model objects: Click on an object and go to the drawing.
- Can show the Workface Planners the **current state of any drawing**: On Hold, IFR, IFC, Revised, Red lined, As built.
- **Add Documents** to the IWP: Allow the Workface Planners to add, remove or update documents directly from the document control database or scanned images.
- Be able to show the Workface Planners the **current state of any material** component: Fabricated, Shipped, Received, Issued, Installed, Revised.
- Develop a material list for each IWP: Create an electronic **list of materials** in Excel for import into material management systems.
- Can sequence and **schedule IWPs**: display start dates for IWPs
- Play a **4D simulation** of either CWPs or IWPs: Allow the Workface Planners to simulate the sequential start of construction activities in a 3D environment.
- Generate **progress scorecards**: A list of the components with the planned value and the rules of progress.
- **Receive physical progress**: Facilitate the entry of progress against an IWP when the work is done:
- Receive **actual hours** burned for each IWP: Facilitate the entry of timesheet hours against each IWP.
- Produce reports that show the **cost performance index** (productivity factor) for each IWP: hours earned against hours burned for each IWP.
- **Print the IWP**: Allow scanned documents and reports to be configured into a user defined format and saved as a PDF.
- Be able to add a single **component to multiple unique plans**: Allow the Workface planners to build task packages for install, connect, test, heat trace, insulation and turnover.
- **Manage Constraints**: List constraints as a checklist and allow them to be progressed.
- **Display progress**: Generate 3D images that show stages of progress (IFC, fabrication, paint, received, installed, connected, tested, turned over) in different colors with summary information (% complete).
- Generate **Planned Value for groups of objects**: Allow the Workface Planners to generate a material take off for any CWP or group of objects and to then calculate the planned value.
- Be **simple** to use: Be intuitive and allow the Workface Planner to be functional without intensive training.

Our experience so far has been that there are only two software products that can do all of these tasks: Hexagon's **Smart Construction** and Bentley's **ConstructSim**. We have successfully applied both to mega projects and the set-up cycle and criteria for information is very similar. The functionality of the end product has much more to do with the data inputs and the way that they are utilized, than the software. While I am confident that there will be other software products in the future that can support the needs of Workface Planning, right now I don't know of any other products that cover all of these basic requirements.

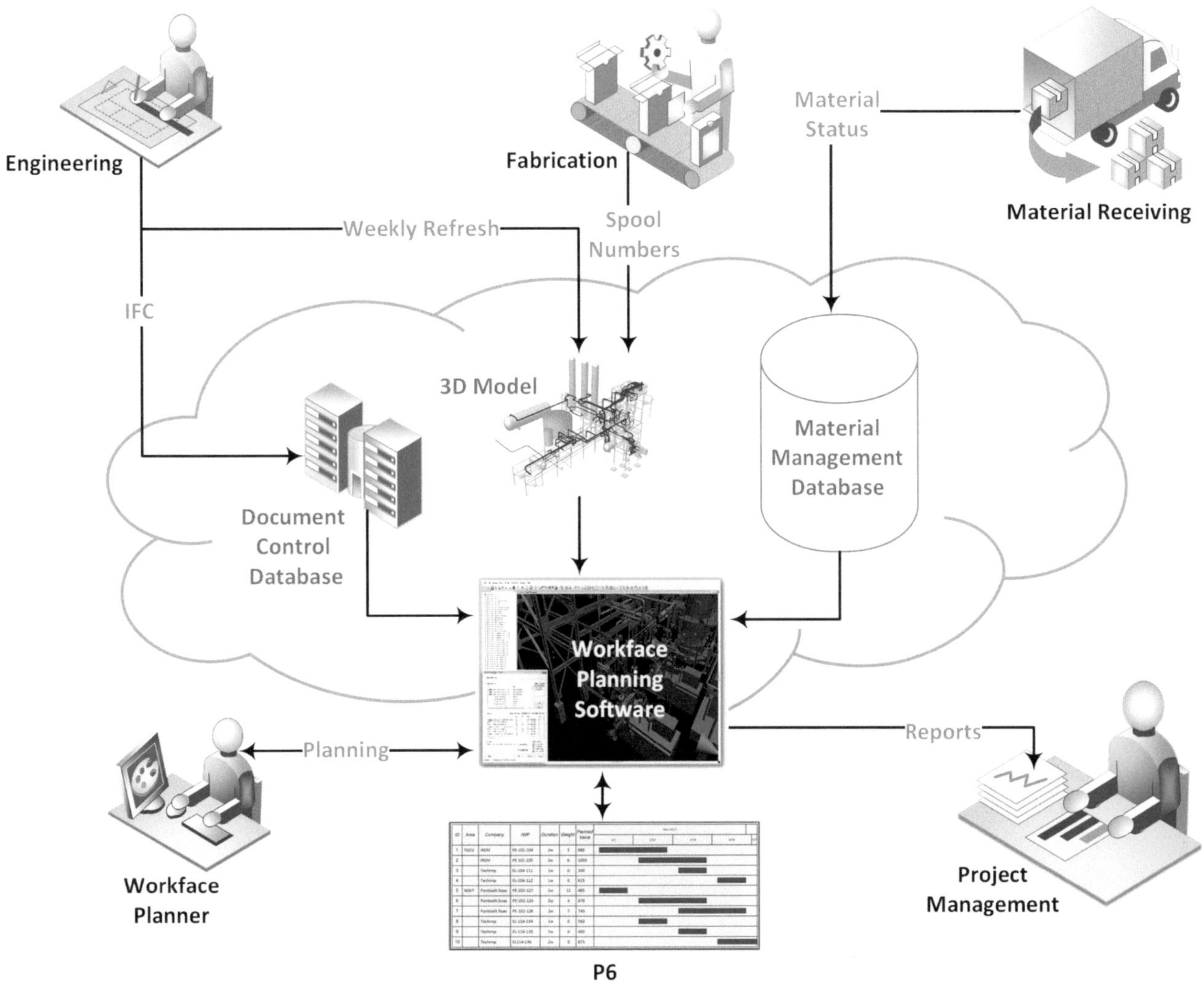

When you get to the stage where the WFP software is established on the Cloud and it is integrated with document control, material management and the project schedule you will have a single source for most of your project information. Normally when our projects reach this level of development the Workface Planners become the key portal for questions and answers. Then as the project management team get use to the idea that all the data is directly available to them on their laptop or phone, they start to mine data themselves directly from the WFP software.

The Information Manager is directly responsible to design this overall system, establish the project cloud and facilitate a proof of concept between each of the interfaces. They are then responsible to stand up the WFP software, populate it with all of the relevant data and maintain its operation for the entire project.

D. 3D Model Attributes:

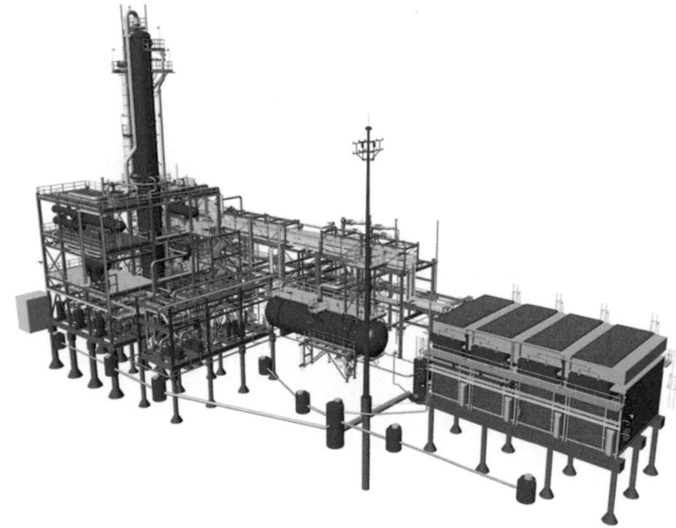

The layout of design areas in the 3D model and the structure and content of the attributes for each component are vitally important to the functionality of the WFP software. Our model for how we ended up with an answer to this very complex question was to start with a list of questions that we anticipate the project will need to know. Then we aligned the attributes to be able to answer the questions. Think of the 3D model as a database that has a comprehensive list of the project components and each component has a list of features (attributes). Then when we are mining data we can ask how many of this item has this unique feature (CWP, material received, erected, turnover system). Then along with the answer we also get a picture that shows the components. Now think about what sort of complex questions that you would like to answer when the project is 25%, 60%, 98% done and that will tell you which attributes that you need.

Even before that we had to start with the actual structure of the model so let's start with the design areas (small divisions in the 3D model): The key for DAs is that they must be developed to fit into the model for Construction Work Packages that were identified during the Path of Construction Workshop. They can be 'many to one': Multiple DAs = One CWP but they cannot be 'one to many'. Beyond that they will also need to be small enough to allow multiple engineering work fronts in a single CWP. Their design will also need to support the possibility that construction will ask for a CWP to be split, so smaller is better.

The way that we start our conversation around attributes is to develop this matrix of 'Wants and Needs', which is then presented to engineering in the early stages of FEED. Then we use this as the foundation for an engineering data workshop where people, who are way smarter than me, talk about whether I can have my wish list or not.

This is the right time to draw upon the resources from the WFP software companies. They have a very clear understanding of the software needs and the Engineering team have a very good understanding of what the model can produce. The middle ground between these two points is the sweet spot for model attribution.

KEY ATTRIBUTES				COMMODITY								
Priority	Attribute	Division Of Responsibility	Source	Civil	Piling	Concrete	Steel	Equip	Pipe	Elec Equip	Cable	Inst
Must	Unique Tag #	Engineering	Engineering	**Y**	**Y**	**Y**	**Y**	**Y**	**Y**	**Y**	**Y**	**Y**
Must	Piece Mark #	Engineering	Engineering	**N/A**	**N/A**	**N/A**	**Y**	**N/A**	**N/A**	**N/A**	**N/A**	**N/A**
Must	Spool #	Workface Planning	Fabrication	**N/A**	**N/A**	**N/A**	**N/A**	**N/A**	**Y**	**N/A**	**N/A**	**N/A**
Must	Component type	Engineering	Engineering	**N/A**	**Y**	**N/A**	**Y**	**Y**	**Y**	**Y**	**Y**	**Y**
Must	Weight (Design Qty)	Engineering	Engineering	**N/A**	**Y**	**N/A**	**Y**	**Y**	**N/A**	**N/A**	**N/A**	**N/A**
Must	Length (Design Qty)	Engineering	Engineering	**N/A**	**Y**	**N/A**	**Y**	**N/A**	**Y**	**N/A**	**Y**	**N/A**
Must	Volume	Engineering	Engineering	**Y**	**N/A**	**Y**	**N/A**	**N/A**	**N/A**	**N/A**	**N/A**	**N/A**
Must	Class (Spec)	Engineering	Engineering	**N/A**	**Y**	**Y**	**Y**	**Y**	**Y**	**Y**	**Y**	**Y**
Must	Diameter	Engineering	Engineering	**N/A**	**Y**	**N/A**	**N/A**	**N/A**	**Y**	**N/A**	**Y**	**N/A**
Must	Wall Thickness	Engineering	Engineering	**N/A**	**Y**	**N/A**	**N/A**	**N/A**	**Y**	**N/A**	**N/A**	**N/A**
Secondary	Service	Engineering	Engineering	**N/A**	**N/A**	**N/A**	**N/A**	**N/A**	**Y**	**N/A**	**N/A**	**N/A**
Must	Insulation	Engineering	Engineering	**N/A**	**N/A**	**N/A**	**N/A**	**Y**	**Y**	**Y**	**Y**	**Y**
Must	Fireproof	Engineering	Engineering	**N/A**	**N/A**	**N/A**	**Y**	**N/A**	**N/A**	**Y**	**Y**	**N/A**
Must	Heat Trace	Engineering	Engineering	**N/A**	**N/A**	**N/A**	**N/A**	**Y**	**Y**	**Y**	**N/A**	**Y**
Must	On/Off Module	Engineering	Engineering	**N/A**	**N/A**	**N/A**	**Y**	**Y**	**Y**	**Y**	**Y**	**Y**
Must	Module #	Engineering	Engineering	**N/A**	**N/A**	**N/A**	**Y**	**Y**	**Y**	**Y**	**Y**	**Y**
Must	CWA	Engineering	Construction	**Y**	**Y**	**Y**	**Y**	**Y**	**Y**	**Y**	**Y**	**Y**
Must	EWP	Engineering	Engineering	**Y**	**Y**	**Y**	**Y**	**Y**	**Y**	**Y**	**Y**	**Y**
Must	CWP	Construction	Construction	**Y**	**Y**	**Y**	**Y**	**Y**	**Y**	**Y**	**Y**	**Y**
Must	IWP	Workface Planning	Construction	**Y**	**Y**	**Y**	**Y**	**Y**	**Y**	**Y**	**Y**	**Y**
Secondary	WBS	Engineering	Project Controls	**Y**	**Y**	**Y**	**Y**	**Y**	**Y**	**Y**	**Y**	**Y**
Secondary	Material Type	Engineering	Procurement	**Y**	**Y**	**Y**	**Y**	**N/A**	**Y**	**N/A**	**N/A**	**N/A**
Must	Material Stock Code	Engineering	Procurement	**Y**	**Y**	**Y**	**Y**	**N/A**	**Y**	**N/A**	**Y**	**N/A**
Must	Design Drawing	Engineering	Engineering	**Y**		**Y**	**Y**	**Y**	**Y**	**Y**	**Y**	**YN**
Must	Fabrication Drawing	Fabricator	Fabricator	**N/A**	**N/A**	**N/A**	**Y**	**Y**	**Y**	**Y**	**N/A**	**N/A**
Must	P&ID	Engineering	Engineering	**N/A**	**N/A**	**N/A**	**N/A**	**Y**	**Y**	**Y**	**N/A**	**Y**
Secondary	General Arrangement	Engineering	Engineering	**N/A**	**Y**	**N/A**	**N/A**	**Y**	**Y**	**Y**	**N/A**	**Y**
Secondary	Connection detail	Fabricator	Fabricator	**N/A**	**Y**	**N/A**	**Y**	**Y**	**Y**	**Y**	**N/A**	**Y**
Secondary	RFID#/Bar code	Fabricator	Procurement	**N/A**	**N/A**	**N/A**	**Y**	**Y**	**Y**	**Y**	**Y**	**Y**
Must	Engineering System #	Engineering	Engineering	**N/A**	**N/A**	**N/A**	**Y**	**Y**	**Y**	**Y**	**Y**	**Y**
Must	Turnover system #	Engineering	Operations	**Y**	**Y**	**Y**	**Y**	**Y**	**Y**	**Y**	**Y**	**Y**
Secondary	Activity ID	Workface Planning	Project Controls	**Y**	**Y**	**Y**	**Y**	**Y**	**Y**	**Y**	**Y**	**Y**
Secondary	PC Cost code	Engineering	Project Controls	**Y**	**Y**	**Y**	**Y**	**Y**	**Y**	**Y**	**Y**	**Y**
Secondary	Weld Number	Engineering	Engineering	**N/A**	**N/A**	**N/A**	**N/A**	**N/A**	**Y**	**N/A**	**N/A**	**N/A**
Secondary	Bolt up Number	Engineering	Engineering	**N/A**	**N/A**	**N/A**	**N/A**	**N/A**	**Y**	**N/A**	**N/A**	**N/A**
Must	Inspection Reqs	Engineering	Engineering	**N/A**	**N/A**	**N/A**	**Y**	**Y**	**Y**	**Y**	**N/A**	**Y**

E. Document Control:

The ideal model for Document Control is a system where Engineering can issue documents to a central database, tell everybody that they did it (transmittals) and then let the project stakeholders help themselves as they need to. The database can be updated with revisions in real time and it will store redline drawings that show field revisions. Essentially creating a single source of project documents that is kept current with a zero lag between the creation of documents and their availability to the contractors.

Now we add to this the revolutionary idea that we could link a 3D model to that database and you could click on an object in the model and go directly to the latest revision drawing.

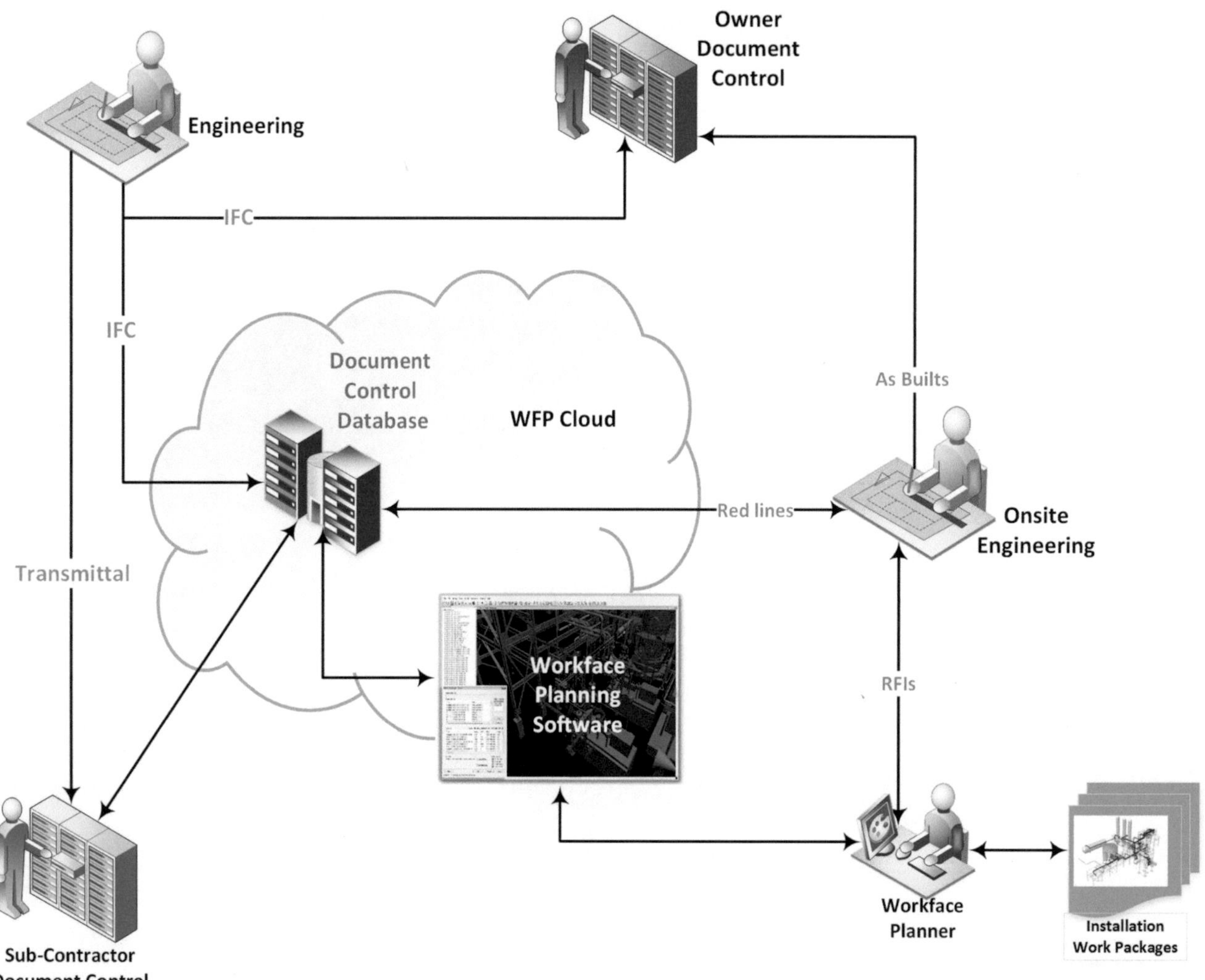

This is what you will get when you set up WFP software and host the document database on the WFP cloud. It's simple, it operates in real time, it absolutely minimizes the administration that is required and it creates a single, common, version of the truth.

The contractors normally don't trust the idea of a virtual supply of documents where they can have electronic access without physically receiving them, (once bitten twice shy), so we usually allow them to operate in parallel with this system. Printing documents from the cloud as they are released and storing the hard copies in their own Document Control system. It doesn't take too long for them to realize that the Foremen get all the documents that they need through the IWPs and if anybody else wants a document that the easiest way is to go see a Workface Planner or to use their read only access through the WFP software. After idly sitting around for a month or so, the Document Control Clerks are then normally moved on to support other areas. As part of the process of making change as easy as possible, you can let this run its natural course.

The application of a system like this where you set up your document control in an environment where the project stakeholders can get direct access to the motherload of project documents, is what we think of as low hanging fruit, in the struggle to optimize projects.

The setup is just a months' work for a couple of IM people, you can lease a cloud environment for a couple of thousand dollars per month and it is the perfect environment for unfettered access to project information.

When you make a statement like that the question of security will come up and our standard answer is that the Central Intelligence Agency (CIA) is now based on the cloud. It does take some design and effort by the Information Manager to set up the protocols that will manage the stakeholder interfaces, but security and backup features are readily available.

All tolled, the project cloud is a small investment that will have a significant positive impact on the whole project.

To fully understand how much difference this makes, it helps if you have worked on a project where the document control systems evolved from hard copies delivered in shoe boxes, managed by an army of crotchety old people, where you had to bribe your way into the field office and stand patiently in line for outdated documents so you could build things that you knew were going to be pulled apart. Don't laugh, this is real and is probably the current system being applied to a project near you.

Two sided Isos:

While we are talking about document control, and because I'm not sure which chapter this belongs to, let's spend a moment on two sided isos. While it is not really part of AWP the positive impact of two sided isos on the communication of ideas is far too important to leave out.

The idea is simple: on the back of each Isometric drawing we ask the engineering team to print a 3D image of the pipe from the front page. So now the end user can get the detailed information from the front page and the orientation of the pipe from the back page.

There are probably instances of where this idea was used over the last 20 years, but it really came to light when the Research Team 327 from CII presented the idea at their annual conference in 2015. The team, led by John Fish, conducted some studies where teams were given ordinary isos and other teams had the two-sided isos. The field trials conducted in 7 different regions using 57 pipefitters, constructed plastic pipe models. The results showed that the two-sided iso teams had a 16% faster build time than the traditional one-sided iso teams.

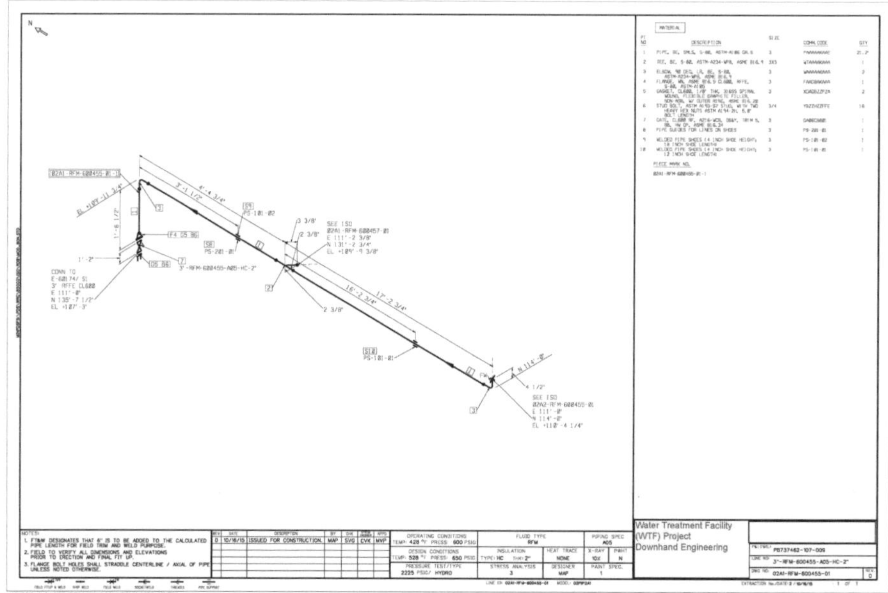

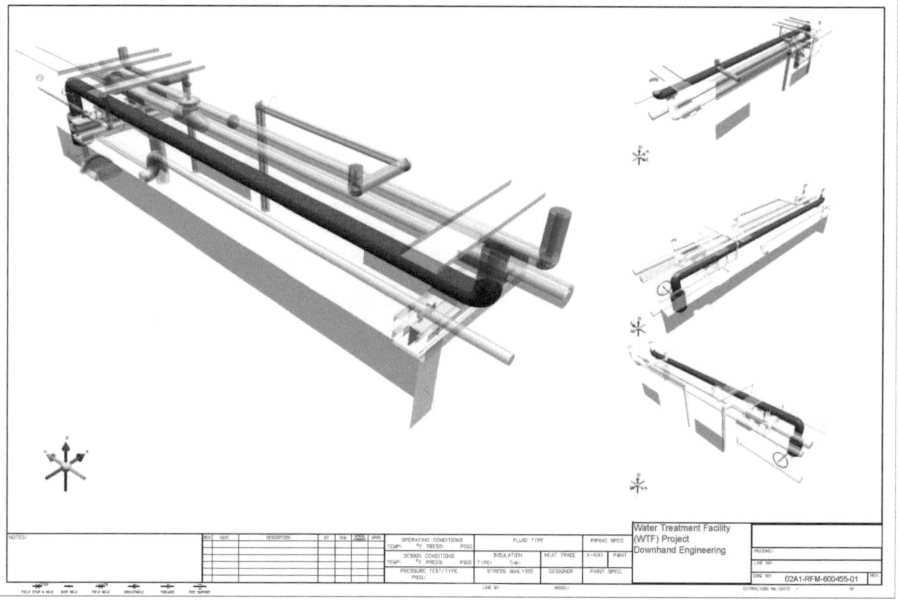

While this number has some significance, for me it is more about 'why not'. On the projects where we have instituted this as a standard, the engineering teams tell me that it is a simple programing issue and doesn't require much effort at all. The benefit to construction cannot be overstated in terms of communication. On top of the obvious benefits, two of our recent projects have been in countries where English was the second or third language, so this pictorial form of communication is almost mandatory.

I included this real-world example because it also shows that two sided isos are an effective way to quality check the model. There is an error in this drawing (the Tee is missing from the model shot) that was not detected until we produced this side by side comparison.

F. Procurement Information

One of the issues that we have with the idea that the WFP software should be the mothership of project information and that the key source of data should be the engineering 3D model, is that information also comes from other places.

Fabrication is a good example of this. Depending upon how the Engineering team like to execute their scope you may find that the 3D engineering model has no spool numbers or steel piece mark numbers, these are often added by the fabricators.

The trouble for us is that if the Workface Planners cannot select a unique spool or steel member in the 3D model then it is impossible to plan it, order it or progress it. So, getting the unique identifiers from the fabricators and getting them into the 3D model is a fundamental requirement.

So then how do we get the information? Let's start by asking for it.

The way that we ask for it is that we stipulate the requirement in the contracts with enough detail to tell the fabricators that we expect the data in an electronic format on a weekly schedule. The response from the fabricators is often that they don't have the information, which means that they have it, but they don't know how to find it, extract it or deliver it, or that they can't be bothered. Either way you may have to send the IM team into the shop to show them how to extract and send the data.

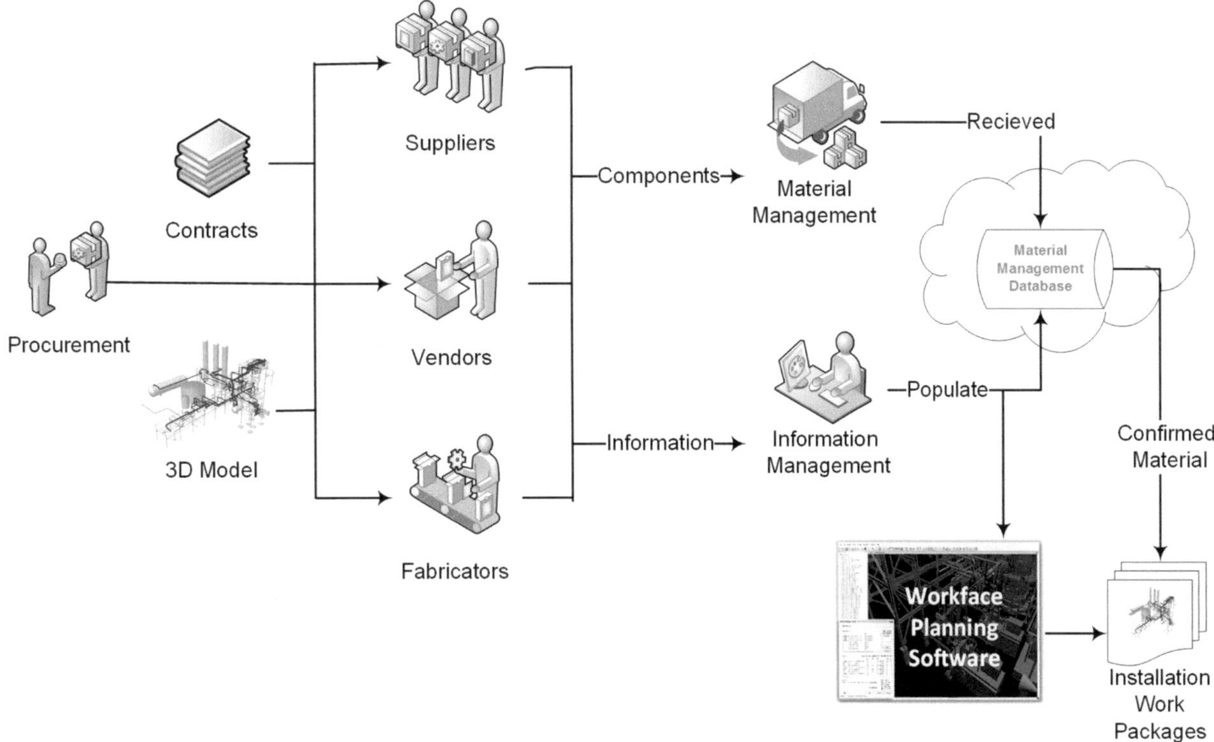

Not getting the electronic data is not an option, it must be identified as a deliverable and is as important as the steel member or the spool. Having components in the yard is of no value if we don't know that we have them. Raise your hand if you have recreated a spool that you couldn't find only to find one that looks just like it in the pile of stuff left over after the project. We had a recent project where this was not a spool, it was a module, which is still just a line item on a spreadsheet and easy to lose when you have hundreds of them.

So data is King, to be treasured and protected at all costs (Like the King in a chess game) and the right information on the right screen at the right time is the key to a successful project.

It should be noted that there are instances where we have found engineering companies who generate their own piece mark numbers and who run spool generation software that assigns spool numbers. This means that the information is in the data when we send it to the fabricators, which works well, however we have also found that the fabricators like to make small changes for different reasons so you will still need a way to capture the final numbers.

Material Tracking:

If you have worked on a project site, in an office building or gone to school in the last ten years you probably used an RFID badge that you swiped on a scanner as you entered a control zone. The same technology is being used more often now to track materials.

In my limited understanding, there are three choices for tracking materials:

- Active RFID, which emits a signal that can be picked up within a set range
- Passive RFID, which operates like your ID badge against a proximity reader
- Barcode, which needs a hand-held scanner to read the barcode

If you want to understand what you have onsite and you don't want to burn thousands of hours on data entry then you will need to start thinking about using one of these systems.

My own thoughts are that your choice should be related to complexity.

For projects up to $1Billion, if you have an established material management database and an organized yard then you can manage quite well with barcodes or passive RFIDs

If your project is over $1Billion, spread over many sites or has laydown yards spread across a wide area then active RFIDs will allow you to find and track material that your system lost.

Either way you will need to engage software that translates the barcode or scanned data into a line item in your material management database. If you do choose to use barcodes, then you will find that most of the fabrication shops already use them and that you can have influence on the way the tag is laid out. This allows you to have your own nomenclature printed on the tag so that your construction team can read the spool or piece mark number when it is in the field.

Pipe Spool Tag

PIECE MARK NO.

2141-SC-0699-03-2

Structural

PIECE MARK NO.

BEAM86780-2

We will spend some more time on material management in the Workface Planning chapter, for the purpose of Information Management you need to know that whatever your decision is on material tracking, it starts at the fabricator. The deliverable is an information management system that support the idea that we need every component in the 3D model identified by its unique name. This means that the contracts for fabrication need to spell out exactly what needs to be installed on the tags and then how often and in which format the information is to be transmitted to the project.

ISO 15926

We talked a little bit about ISO 15926 in the original book, Schedule for Sale, and the principle is still the same. Wikipedia has a good explanation of the standard and FIATECH have lots of documents on it for those of you who need more detail. The principle in my head is that every component (pump) has a requirement for say 22 pieces of information and that when the 22 boxes of information have been answered and the data is in XML format then the information is compliant with the standard.

Sounds simple enough, however, getting the standard established as an industry requirement is having trouble getting traction. The idea that we should have industry standards for component information, across the whole industry, will eventually be the rule and here is where you can help. FIATECH have done the hard work of developing the standard so that all we have to do is to ask our vendors to supply data with their components that is compliant with ISO 15926. This means that the data is delivered in an XML format so that it can be entered into software that is also 15926 compliant. And if we all start to ask for it then the vendors and software developers will all start to step up to the requirement.

This is very useful when we start to explore the idea that information generated on our projects could be used to support life cycle requirements for the stuff that we build. We already have clients who are asking us to keep 3D models updated with field changes as they are discovered, so that we can turn over the 3D model as part of the turnover process. This is so that they can use it for operation simulations and possibly maintenance and retrofits sometime in the future. This idea of life cycle information utilization brings high value to the client and a focus on the quality of information that we initially collect.

H. Cost Codes

Now that we have explored the rules for the generators of information through the 3D model, document control and fabrication we need to start to think about how we will make use of the data that will be generated by the project, through cost codes.

In my mind project controls has two primary functions (neither of which is control). We use the WBS process to create bite sized portions of the project that are quantitatively applied with data to show a projection of how long the project might take and how many resources we will need. Then we string these together in a sequence that make sense to construction and we end up with a schedule. This forms a projection, like a crystal ball for the project. Years and years of refinement have made us pretty good at building these projections, but it's still a guess, and sometimes it's a WAG.

Then we have the stuff at the other end, the actual results of our efforts, which is formulated through the use of cost codes, that we compare to the crystal ball in some vain effort to prove that we were right and that we can actually see the future. But alas, this process normally proves the exact opposite, we are mere mortals and the best we can hope for is to be able to keep a tally of what we did against what has yet to be done.

This doesn't mean that you have an excuse to not build a schedule, we still need to utilize our gift of the frontal lobe to build a schedule so that we can get as close as possible to the crystal ball. It really just means that you should not bet the house on the outcome.

The highest value function of project controls comes into focus when we use actual progress to track our current performance trends and use that to help make project decisions. It's a bit like steering a ship on the ocean, looking past the bow it's hard to tell where you are going, but if you take a look off the stern you can see the trend. These trends show us the impact of actions and decisions, which have given us the general direction of the project, so it is important to make the information as accurate as possible and also to keep the cause and effect cycle as tight as possible.

With this in mind, we need to start thinking about how to arrange the building blocks of data generation so that we can use the information/knowledge as a project management tool.

It typically starts with a conversation between the Information Manager and the Project Controls team, where we lay out the puzzle pieces and create a vision of what the project will need to know to make informed decisions.

We know that we will need a set of installation rates for all disciplines that have been agreed between the construction contractors and the project management team. We will also need a set of progress rules that spell out how much credit we claim for each stage of construction. Then we will need a set of cost codes that will allow timesheets to be tallied against segments of work, so that we can tell how many hours we spent getting a specific portion of work done.

These are the standard seeds required for a healthy crop of project controls reports, however the real actuator that pours sunshine and water on these seeds are the processes that we use to collect the data. They need to be simple for the Foreman and effective in that they collect timely information that is accurate, which is the real hard part. The detailed process of how we develop cost codes that are fit for purpose is covered in section D of Chapter 5.

Unless you are reading this for the first time and it is sometime after 2030 or your project controls people had the miraculous good fortune to have worked on a project where they had exposure to WFP software, your conversation should start with a description of a mythical world where the motherload of project controls data is visualized in a 3D environment on a project cloud. With Planned Value and Earned Value magically tracked against every component and rolled out in dashboard reports from a single press of a button. Then after the euphoria has passed and they are coming back into reality, tell them that it could be true and that they could own such a wonderful tool, but there's a price.

The first thing is that they need to let go of some old customs and think outside the circle. The creation of a system that synchronizes work with time and cost will take a bit of change, faith and courage.

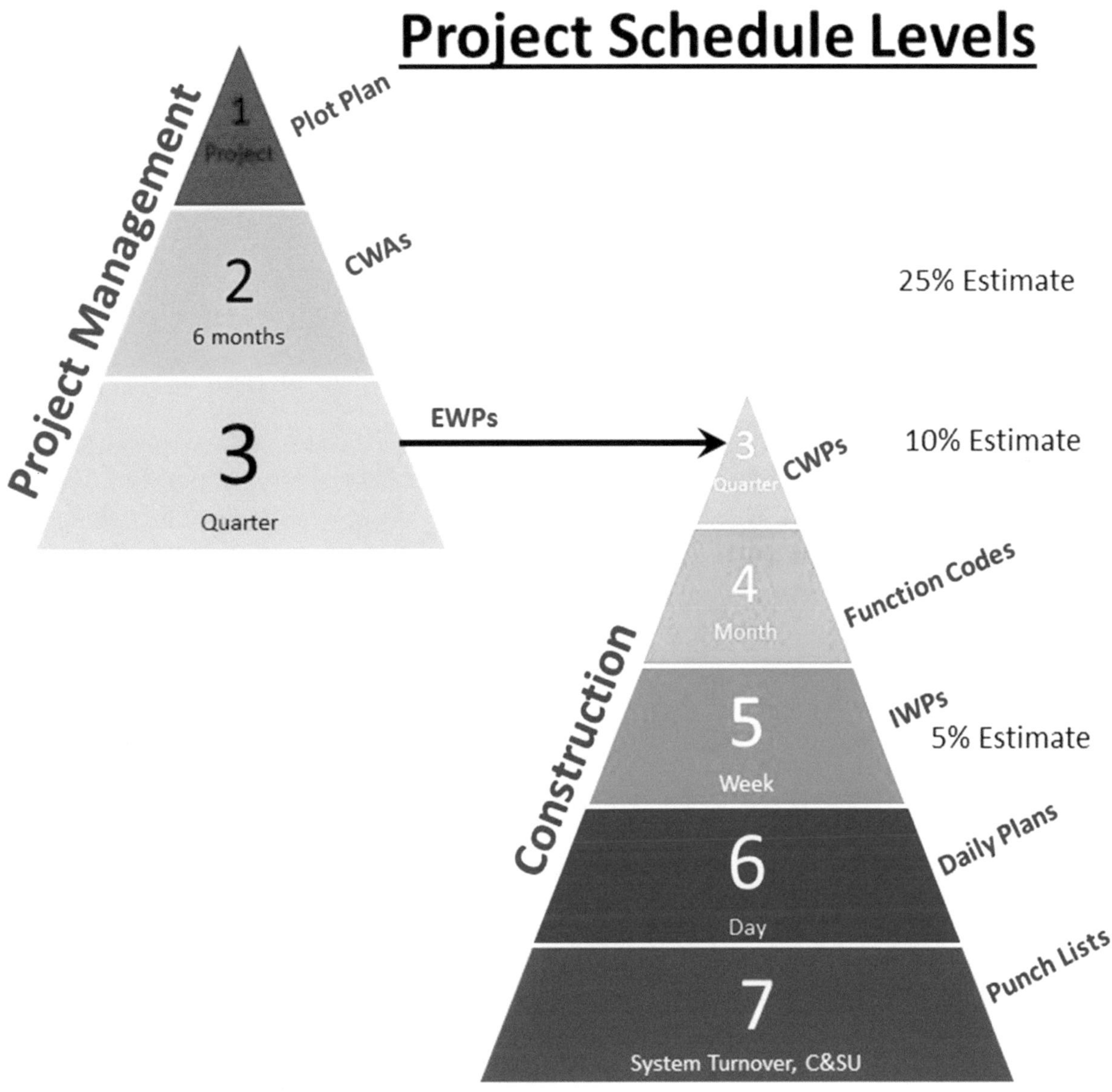

A good place to start is to establish standards for each level of the schedule that also have a structure for the level of estimating confidence required at each level (+/- 10% for CWPs)

The level 3 schedule will then have to sacrifice most of it's details and be rolled up to CWPs. We will only have one level 3 schedule which we will share with the construction contractors and they will develop IWPs for their level 5 schedule and the schedule activities will be named by their WBS code.

Cost codes will have to be developed to support the work breakdown structure, not the cost breakdown structure (although it could do both). With codes for IWPs that can be rolled up to CWPs and there will be no codes for large bore, small bore, heavy steel or light steel. (that's not the way that it gets built) Timesheets will be cost coded with the IWP number and be used to collect delay codes.

The installation rates used to track earned value must be the same ones used to generate the project estimate and the rules of progress must be agreed by the contractors.

Project controls reports will be 'on demand' for the stakeholders, accessing the cloud from their desktops or phones, creating a world where live data is mined from a single source of project information. This will create an environment where the message cannot be managed, which also means that we don't need most of the project controls people that we currently have.

(You may want to let this last message leak out naturally, over time.)

If you can get all of this on the table and not get kicked out of the PC office then you will be well on the way towards establishing WFP software as the source of all truth and the single greatest revolution to take place in project controls since the mapping of critical path by Morgan Walker of Dupont in 1956.

The output from this alignment of rules with Project Controls will be a very simple level 3 schedule that only shows EWPs, PWPs, CWPs and major equipment, along with a set of installation rates and rules of progress that can be added to the WFP software and finally a set of cost codes that look a lot like the name of IWPs and schedule activities.

Work = Time = Cost.

So now it is approaching the end of the week, a week where you have done all of the right things and put in some serious effort and overtime to make sure that you have the foundation laid for a successful outcome. Well congratulations, this is payday.

Chapter 7: Workface Planning

I often get asked at the start of projects, after I have explained the process and outcome to the project management team and they have agreed to march down this path, "Is there anything that I need from them" (before we get started). The answer that I am using more often now is that I need the project management team and our sponsor to 'watch our back'. That usually draws a raised eyebrow and quizzical look. Then I let them know that there will come a time on the project where we have successfully delayed engineering by demanding EWPs that support CWPs, driven up the cost of procurement through contracts that prioritize sequence over volume, alienated the entire project controls team by daring to rock their boat with simplified schedules and Earned Value management and then created chaos in the construction office by only letting them start work that can be finished, all without producing any tangible results. It is then that we need the project to have the confidence to 'stay the course' and guard our back.

The start of construction is that time, and it will get worse before it gets better.

It helps if you can disguise any doubts that you have and make sure that you don't emit the smell of fear, put your best foot forward and make it look like you have done this a hundred times.

The real pucker point comes at around 15% construction complete, when things are going sideways because your field guys aren't executing the plan and you have delays caused by things that you could not have imagined. Fear not, the foundation that you laid will produce fruit and you will eventually see a rhythm start to flow between planning and execution. However, you do still have a motherload of hard work to do:

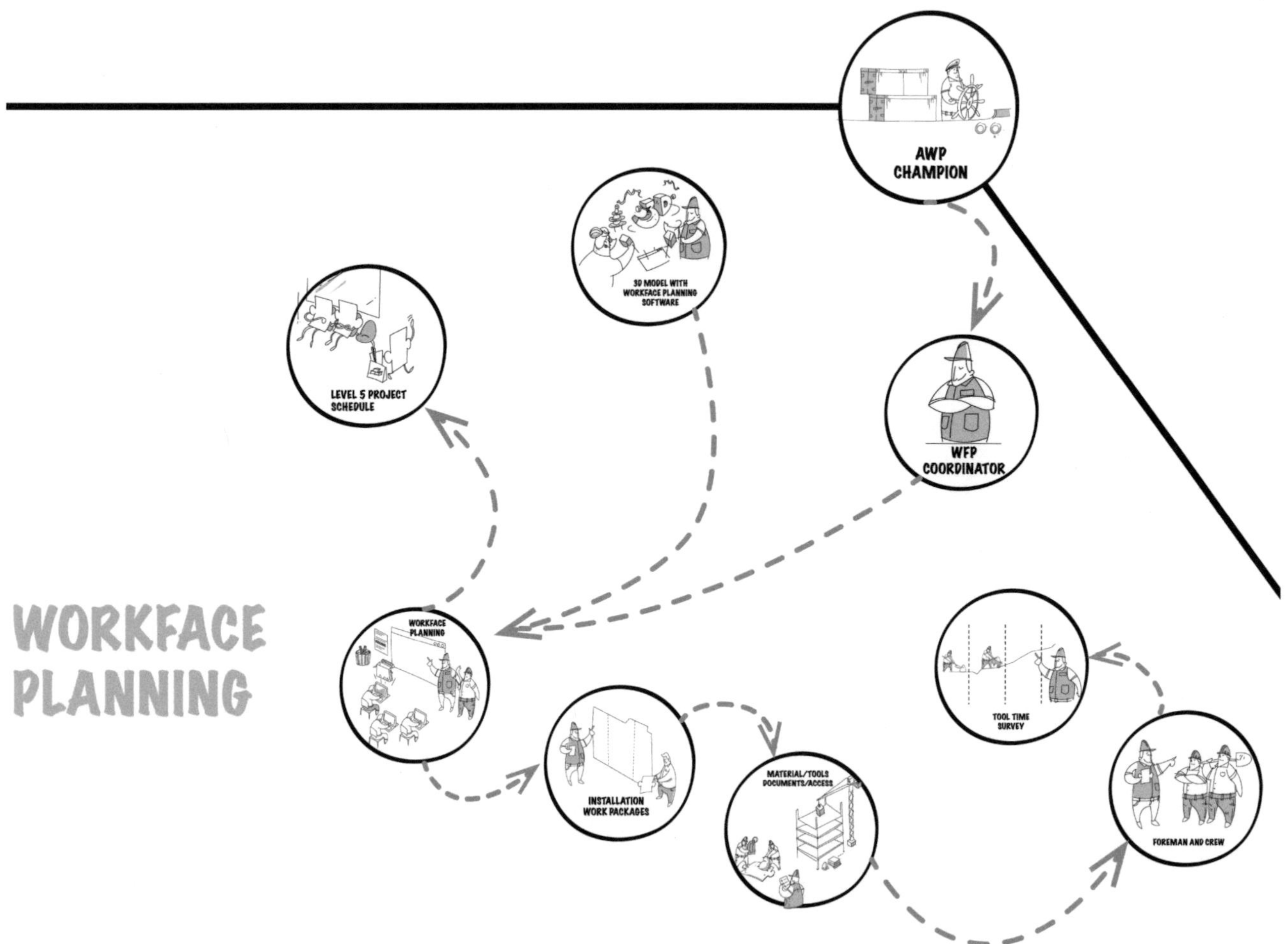

If you had the good fortune to read the first book: Schedule for Sale, then you will have an effective foundation for this section and you will notice that we don't spend as much time on the details of support services: Material management, document control, scaffold management and equipment management. We will touch on them and some others, but the focus of Workface Planning in an Advanced Work Packaging environment is about getting and executing Construction Work Packages, so that is the primary theme.

The AWP quick start guide has the details of IWPs and constraint management.

Following the AWP infographic, we will explore:

A. Construction Work Packages
B. Workface Planning Coordinator
C. Installation Work Packages and Constraint Management
D. Workface Planning Software
E. The Level 5 schedule
F. Field Level Execution
G. Tool Time Studies

A. Construction Work Packages:

One of the key features of the AWP process is that there are gates in the production process that demand compliance with AWP rules: The Engineering team only get credit (and payment) for the last drawing in an EWP and the fabrication team only get credit (and payment) for the last spool, steel member or module for a PWP. This creates a utopia for each CWP that we often hear spoken about as the good old days, where we had all of the drawings and materials before we started construction. The need for Fast Track construction, where we execute work fronts in parallel and as early as possible, often gets the blame for our departure from this fundamental. This method of mini projects distinguished as CWPs gives us a working model for the foundation of documents and materials before we start, in a format that allows us to execute parallel activities in a Fast Track environment, giving us the best of both worlds. High productivity from having what we need and tight construction windows that come from multiple work fronts.

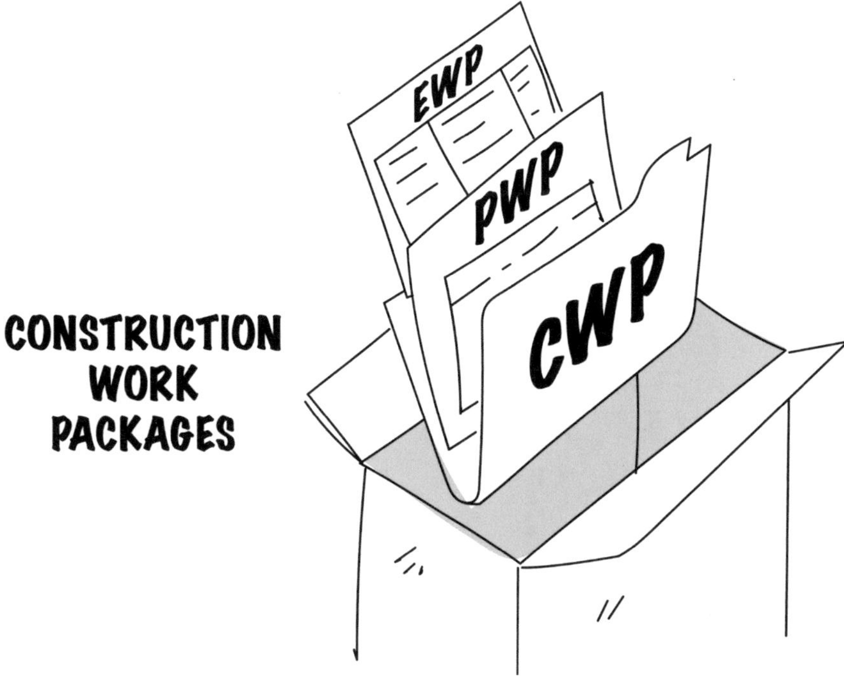

The rationale behind the size and contents of a Construction Work Package (CWP) is that it is a logical association of scope for a single discipline that can be planned by a single Superintendent.

The sweet spot for this point is around 30,000 hours, which is about three months work for 50 people. We often stretch this to 40,000 to accommodate a logical association of work, but anything over that becomes difficult to plan without being subdivided. This same quantity of work will pan out to be about 250 drawings and 10,000 feet of pipe or 1000 tons of steel, which are good batch sizes for engineering and fabrication.

Then, as described in the AWP section, when we get to the point where the scheduled execution date is within the next 90 days, the CWP is pre-released by the CWP Coordinator to the Superintendent for planning which leads into the squad check. The Superintendent develops a plan for how to dissect the scope and the Workface Planner captures the strategy so that they can develop Installation Work Packages after the workshop.

This transition is the start of the Workface Planning process and draws attention to the engineering and procurement deliverables of last drawing for an EWP and last spool, steel member or module in a PWP.

B. Workface Planning Coordinator:

WORKFACE PLANNING COORDINATOR

The placement of the WFP Coordinator from the project management team to be the co-located lead of the construction contractor's WFP team is an idea that takes a bit of getting used to, but we learned the hard way that this works and any other variation doesn't.

It is a nice idea that the construction contractors would know how to execute Workface Planning and just get it done while we watch in awe from the sidelines, but they don't and it is quite a leap to learn it by reading this book or by having a procedure thrust at them. It is much better to give them a resource that will facilitate the learning cycle. Somebody who can use their link to the project management team to make sure that they get the deliverables from engineering and procurement.

To help you get your head around the idea of the WFP Coordinator running the contractor's WFP department, it may help to think of them as a coach, a subject matter expert, imbedded by the Owner to train a group of very capable players into a winning team. Responsible for structure and process, but not content. It is very important that the IWPs are generated by the contractors based upon how their Superintendent will execute the scope (Content). Their expertise in construction execution is why they were chosen, their ability to execute Workface Planning is a skill that can be developed.

This model of integration is a good example of collaboration where the WFP Coordinator brings the authority of the project management team to deal with issues and the construction contractors supply the manpower and construction expertise. We have had a series of projects where this model worked very well, under every type of contract. Yes, even lump-sum.

In our lump-sum models the Owner paid for the Workface Planners and the WFP coordinators over and above the contract price. Their reward was that they got their project on time or ahead of schedule and the contractor made very few claims.

If you still think that placing a WFP Coordinator from the project management team is a bit of an intrusion into the contractor's sand box, then think about how we manage successful Safety and Quality programs. The Project management team work hand in hand with all of the other stakeholders to get the best possible team results. That is how we need to think about productivity.

Poor performance is not a contractor problem, typically contractors will perform as well as we let them and everybody pays the price when the contractor fails. If our goal is to get the best possible project results then integrated teams and holistic project thinking is the right way to get there.

If you want to avoid the pain of change and don't want to hurt anybody's feelings, in exchange for poor project results then we can stick with the existing model where we set unrealistic expectations and then sit back and jeer as the construction teams fail. But we already know what those results look like.

When you do get your head around the idea of fully integrated teams and holistic project accountability, this is what your Workface Planning team may look like.

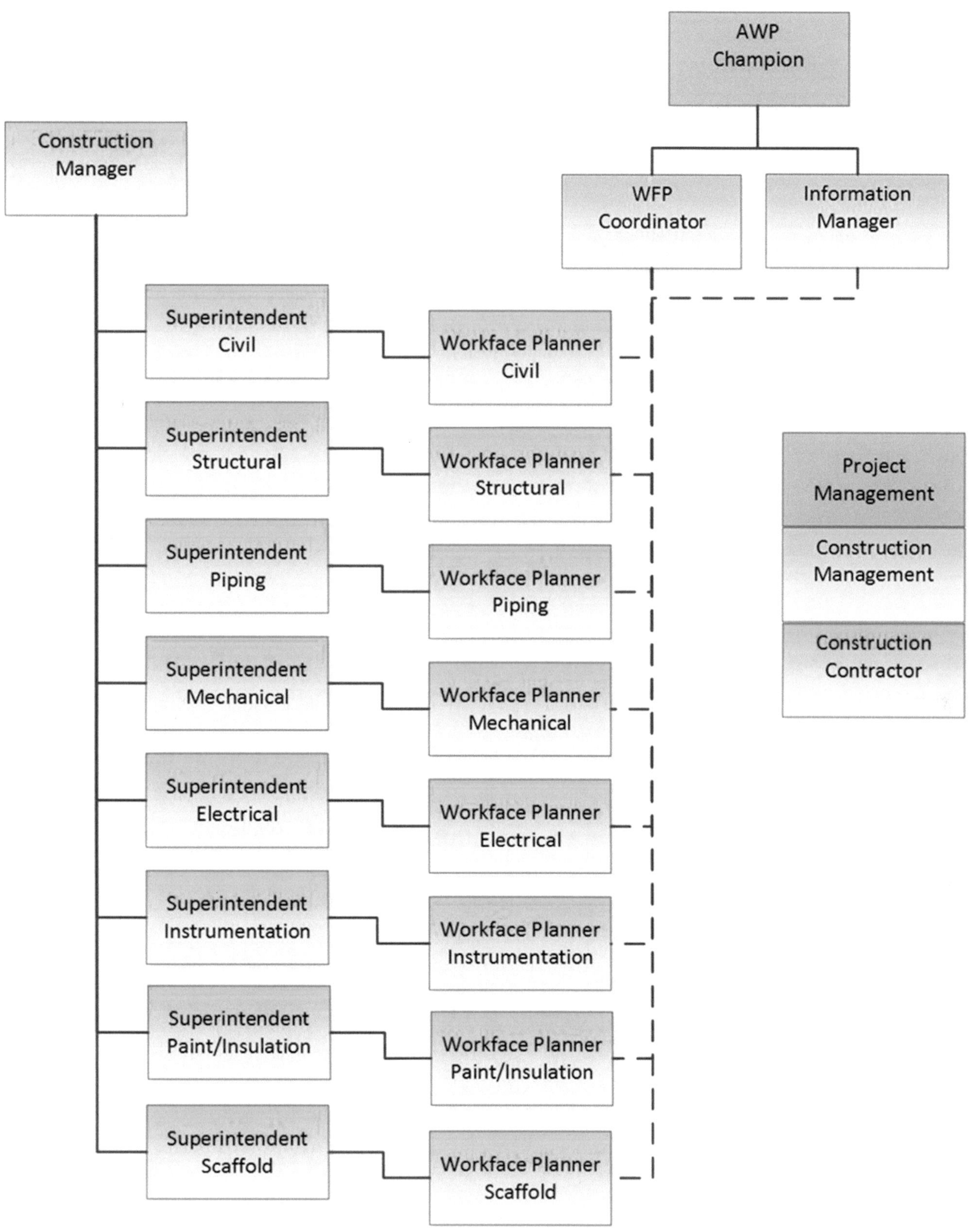

In this integrated model, the WFP Coordinator is responsible to establish the department by working with the construction contractor to select WFP candidates that have the right attitude, skill set and construction experience. The Workface Planners report directly to their Superintendent and then to the WFP Coordinator in a matrix environment. It is very important that the Superintendent sees the Workface Planner as their asset. A major part of the WFP Coordinator's role is to establish this relationship and then to train the Superintendents on how to use their Planner.

The Information manager is also part of this team and makes the transition from project management to construction management by joining this department. They are responsible to stand up the WFP software, deliver the initial training and then provide the day to day support that brings the software to life as a Workface Planning tool.

With the department established the role of the WFP Coordinator then transitions into a facilitator, working to establish standards for Installation Work Packages (IWPs) and constraint management.

Workface Planners:

We have spent a lot of time going backwards and forwards on the description for the ideal Workface Planner, we typically insist that they are trades people with supervision experience but we are given office jockeys or engineering students. While these people are well meaning and quite often do manage to get IWPs built, they still don't have the field experience to build packages that make sense to the foremen.

If your department is big enough, you can hide some construction incompetence and even get some benefit from engineering know how, but 80% of your planners still need to come from the construction trades. You can learn this the easy way or the hard way, but you will learn it. The easy way is to take our advice, the hard way is to employ non-construction people to be Workface Planners and then find out half way through the project that the Foremen only use the IWPs for drawings and 3D pictures, which means that you left half of the benefit on the table, just trying to prove yourself right.

The second most important attribute after construction know how is a get-r-done attitude, then a distant third is computer skills. You can teach somebody to use a computer in an office environment, but you can't teach construction skills or attitude from the office.

The other thing that we do know for sure is that well qualified, hardworking people often fail in this role because they have no WFP infrastructure. You can start to address this with industry training and research from the COAA and CII websites, but the execution of AWP and more specifically WFP takes structure that comes from procedures and experience. Hence the need for the WFP Coordinator.

C. Installation Work Packages and Constraint Management:

This is the heart of Workface Planning and two of the key ingredients that we found present in successful construction companies, when Lloyd Rankin and I did the research for the COAA committee all those years ago (2003 – 2005).

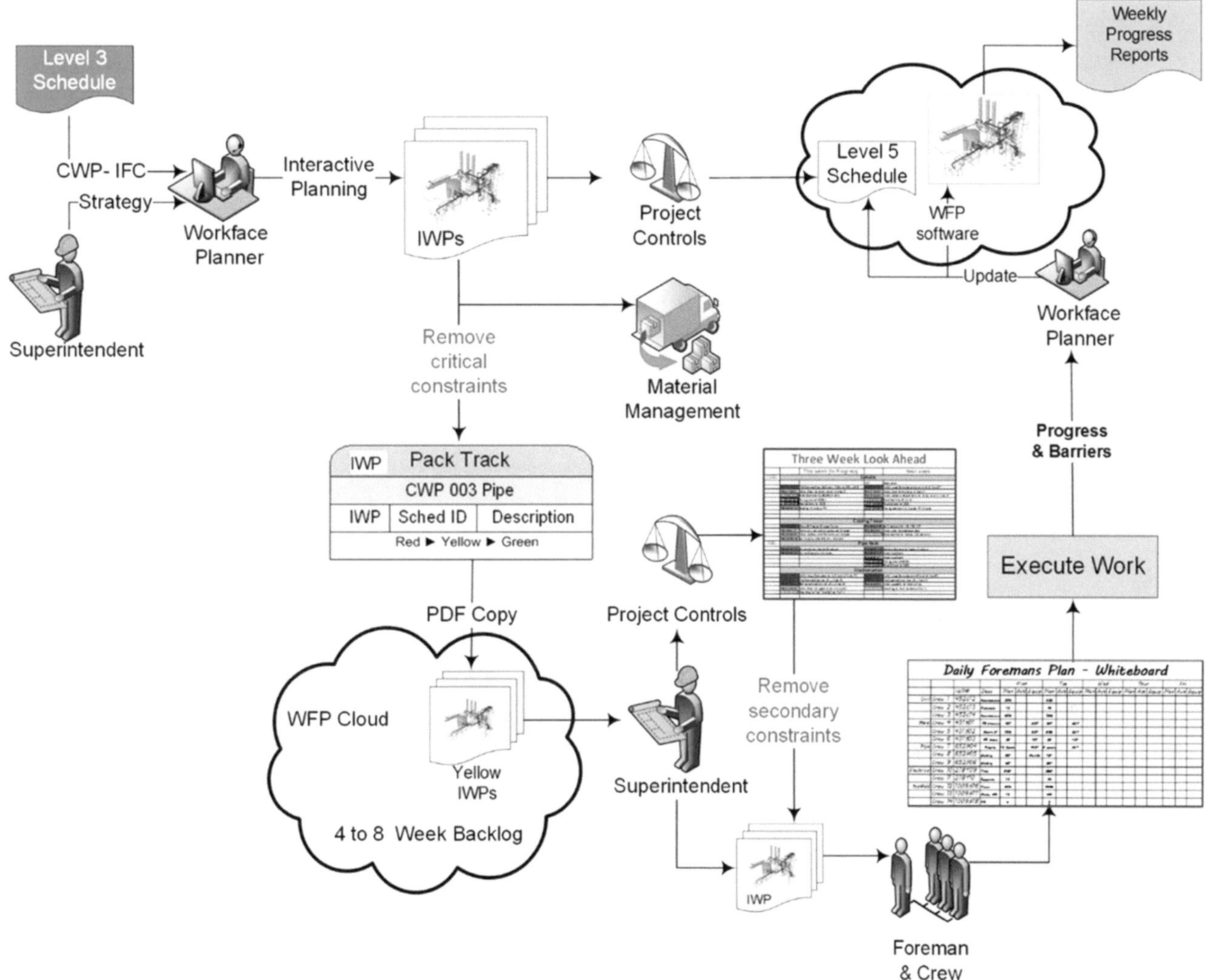

This simplified version of the flow chart shows that the CWP is flagged as being ready for planning as it enters the 90 Day window and after it has been released IFC. The Workface Planner and the Superintendent conduct an interactive planning session (Squad check) with the CWP Coordinator and engineering to flush out the overall strategy and then create the outline for Installation Work Packages (IWPs) and the IWP release plan. The Workface Planner then goes to work in the WFP software and develops IWPs that support the strategy.

Important notes:

> There is a temptation when developing the level 3 schedule before construction has started to add definition below each CWP, if you do that then you need one more step on this flow chart before it begins: 'Throw out any schedule definition below the CWP'. I promise you that no matter how smart you think you are, you cannot define the construction process better than the Superintendent who is going to do the work. If you do, they will ignore it anyway and do it their own way. So save yourself some time and effort: Don't plan below the CWP.

Outsourcing IWPs: The idea of outsourcing IWP creation to the home office or even offshore, has the ring of cost reduction to it and is used by salesmen to show how progressive and 'out of the box' a company is, when bidding on projects. For the same reason just described (Superintendents doing their own thing), this is not practical and 100% of the times that this has been applied the IWPs that were created did not satisfy the field's requirements for planning. Not only do they not support the work, the superintendent has to work around them so they are actually an impediment to progress. This is the wrong place to try to reduce cost. Planners are cheap, crews of welders standing around waiting for something to do, are really, really expensive so your IWPs should be as good as they possibly can be. Besides that, the process of drafting virtual IWPs in the WFP software can be punched out in one or two days for an entire CWP, so why would you have a team of people spend several weeks building IWPs in the home office that get worked around and then abandoned.

Once the virtual IWPs have been created in the WFP software the Workface Planner invites the Superintendent back and plays the 4D simulation of the IWPs. The software pops each IWP 3D image onto the screen in sequence and then colors them to show the different stages of progress. The Superintendent picks up their jaw off the floor and then goes to work fine tuning the content and sequence until they are satisfied.

Now that the Workface Planner has IWPs defined they start the process of constraint management. The first critical constraint is documents, but because we didn't start to plan until the CWP was IFC, we should be OK, with some holds and exceptions that need to be monitored.

The next most critical constraint is material. Workface Planning software has the capacity to display the current status of received material against IWPs so the Workface Planner monitors this and tracks the progress in Pack Track. A simple spreadsheet that shows a matrix of IWPs against a list of constraints.

				90 Day Planning			IWP Assembly				3 Week Look Ahead										
		Weeks prior to execution		12	12	12	4	4	4	4	3	3	3	3	3	2	2	2	1	1	-1
CWP PE3-57	**IWP**	Description	Planned Value	Scoped	IWP Created in 3D	Inserted into L5 Schedule	Documents IFC	Materials Available	Technical Review (RFIs)	Enter Backlog	Enter 3 Week Look Ahead	Bag and Tag Material	Request Scaffold	Request Cranes & Equipment	IWP Hard Copy	Safety	Quality	Resources Confirmed	Preceeding Work Confirmed	Issued to the Field	Work Complete
Civil																					
PE3-57-EW																					
Grade	PE3-57-EW-01	Survey for Grade	840	✓	✓	✓	✓	✓	✓	✓	✓	✓	✓	✓	✓	✓	✓	✓	✓	✓	✓
	PE3-57-EW-02	Strip Top Soil	1340	✓	✓	✓	✓	✓	✓	✓	✓	✓	✓	✓	✓	✓	✓	✓	✓	✓	
	PE3-57-EW-03	Grade to Elevation 1	890	✓	✓	✓	✓	✓	✓	✓	✓	✓	✓	✓	✓	✓	✓	✓			
	PE3-57-EW-04	Grade to Elevation 2	730	✓	✓	✓	✓	✓	✓	✓	✓	✓	✓	✓	✓	✓	✓				
Piling	PE3-57-EW-05	Survey for Piling Placement	620	✓	✓	✓	✓	✓	✓	✓	✓	✓	✓	✓							
	PE3-57-EW-06	Mobilize Piling rig and materials	450	✓	✓	✓	✓	✓	✓	✓	✓	✓	✓								
	PE3-57-EW-07	Install Piles North Side	980	✓	✓	✓	✓	✓	✓	✓	✓	✓	✓								
	PE3-57-EW-08	Install Piles South Side	730	✓	✓	✓	✓	✓	✓	✓	✓										
	PE3-57-EW-09	Cut and Cap Piles North	860	✓	✓	✓	✓	✓	✓	✓	✓										
	PE3-57-EW-10	Cut and Cap Piles South	1250	✓	✓	✓	✓	✓	✓	✓	✓										
PE3-57-CO	PE3-57-CO-01	Survey for form work	820	✓	✓	✓	✓	✓	✓	✓	✓										
Formwork	PE3-57-CO-02	Excavate for form work	1420	✓	✓	✓	✓	✓	✓	✓	✓										
	PE3-57-CO-03	Install form for EB-43	850	✓	✓	✓	✓	✓	✓	✓	✓										
	PE3-57-CO-04	Build Rebar cage EB-43	640	✓	✓	✓	✓	✓	✓	✓	✓										
	PE3-57-CO-05	Construct Forms for CG3-9	790	✓	✓	✓	✓	✓	✓	✓	✓										
Rebar	PE3-57-CO-06	Build Rebar cage CG3-9	550	✓	✓	✓	✓	✓	✓	✓	✓										
	PE3-57-CO-07	Pour EB-43 and CG3-9	350	✓	✓	✓	✓	✓	✓	✓	✓										

If you want to truly optimize the amount of money and time you save, you could apply a 100/80 rule, where no work starts on any CWP until you have received 100% of the engineering and 80% of the materials.

An important note for materials: For many years we had a "Just in time" material mindset where we order materials and have them arrive at the yard just before we need them, however as you know this is a dangerous existence. The idea is to reduce the waste of effort and space in the material management system, but the fragile nature of material delivery promises led us to refer to the process as "Just too late materials". The 5 cents we tried to save in material handling costs many thousands of dollars in construction. This is a good example of why we should not think in silos but should think globally. The optimization of the supply chain must be done with consideration for the overall optimization of the project. Reducing waste in the parts can transfer and multiply the debit to construction.

As each IWP achieves compliance with documents and materials and they have been checked for inconsistencies that would require a Request For Information (RFI), they move from being red to yellow, which means that they have the critical constraints satisfied and that they can be saved as a PDF and submitted to the backlog on the cloud.

We talked about the idea of the cloud earlier, but skipped over the benefit to construction. With all the yellow PDF IWPs stored in the cloud as the backlog and the latest copy of the Three Week Look Ahead posted there, we can give all of the construction people read only access so that they can review the plans and the 'three week look ahead' schedule at any time, from anywhere on any device.

(Normal stuff for the rest of the world, but pretty cool innovation for construction)

D. Workface Planning Software

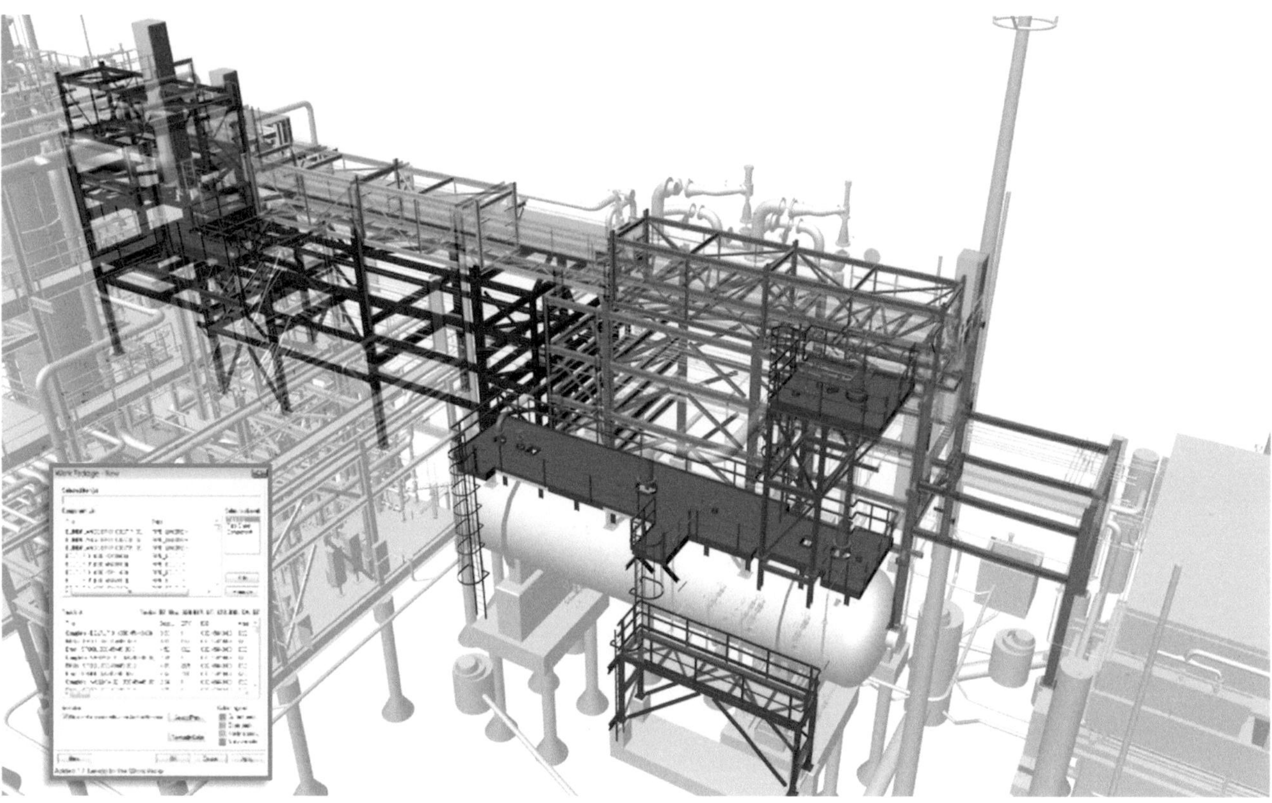

The relationship between the WFP software (Tool) and the process of building IWPs and then managing their constraints (Process) is a good example of the difference between independent and interdependent. Either one could operate without the other but when they are used in conjunction the sum of the whole, is greater than the sum of the parts: 1+1 = 3

We have executed AWP projects without WFP Software and derived great benefit, but there was always this haunting knowledge that we left lots of benefit on the table. The same is true of WFP software applied on projects where the Foreman never received an IWP, there was some benefit but most of the gains were left in the 'didn't get to it' waste basket. As the application of AWP becomes more common I expect that we will stop talking about WFP software as an option and just accept that it is a major part of the answer.

In the Information Management section, we talked about WFP software being a project management tool, not just a tool for creating the IWPs. If this is the case then by the time you get to construction you already have this stood up and it has been churning out reports on iso production and fabrication status for many months. If not then you need to allow yourself 3-6 months of set up time and if you are not sure if it will be 3 or 6, then it is 6.

The key function of the software from this point is to facilitate the creation of IWPs. It does this by allowing the Workface Planners to select a 3D image of a single CWP (3-4 months' work) that has entered the 90-day construction window. Then they click on a single object or a group of objects and add them to a plan. The plan has an hour meter that shows the Planner the Planned Value (PV) of the work and the cumulative total. When the Planner has added enough of the right objects to the plan to make a week of work for a crew, they label the virtual plan based upon the IWP nomenclature. They keep doing this until all of the objects are planned, the result is a series of IWPs (30-60) that represent the Superintendent's vision and strategy. This gives the Planner the outline of the IWPs that will be needed to execute the entire scope of the CWP. The next step is to review the PV for each IWP and adjust for height or other difficulties. Then the Superintendent and the Planner arrange the IWPs into a schedule in excel that satisfies the duration of the CWP, this also creates a basic resource profile that shows human resources, cranes and scaffolds across the 3 or 4 months of construction.

One of the key advantages of hosting the WFP software on the cloud is that it can also be used to store all the project documents. This allows the software to sync objects in the model with documents so that the planner can right click on an object and open the latest revision of the drawing, adding it to the IWP. The same synchronization can be formed with the material management database so that the Bill of Materials for each IWP can be vetted against the materials received onsite. The Planner can then reserve the materials against the IWP and show that the IWP is free of critical constraints (Documents and Materials) and ready to enter the backlog.

Getting the Workface Planners to the point where they can do all of this as easily as I just described it, is the extended role of the Information Manager. Depending on how the Workface Planners are onboarded we may have some group training, but it is usually one on one and over a period of up to 6 months. This is one of the major reasons that we have the WFP Coordinator and the Information Manager imbedded in the contractor's WFP department. The IM can make sure that the software is running at full speed and can teach the planners how to use it in an environment that is tailored to their individual learning style.

At the end of the day a very important part of understanding construction productivity is the ability to measure effort against output. The output will come from earned value calculations generated by the WFP software as the physical progress is entered, the effort needs to come from daily timesheets which are also entered into the WFP software and are cost coded with the IWP number by the Foreman. The hours 'burned' against an IWP will then be divided by the hours 'earned' which will produce a number above 1 if productivity was better than estimated.

As the progress is entered against the IWPs the WFP software changes the status and the color of the components so that along with a summary of the amount of scope that was completed there is also a picture of what was done. While this is a very cool feature that comes from managing scope within a virtual 3D environment, one of the greatest benefits is that it also shows which work-fronts have now opened as a result of the progress.

E. Level 5 Schedule

As the CWP is decomposed into IWPs that have Planned Value and durations we moved into a new level of definition that could be described as having a confidence level of +/- 5%. Thanks to the 4D simulation from the WFP software we also have the sequence of these discrete events, which is the perfect combination needed to create a level 5 construction schedule. An important feature of the Level 5 schedule is that there should be no detail, simply the name of the IWP, (Construct foundation forms, R125) the full IWP with all the details is available to everybody on the cloud.

When you add these events (IWPs) to the project schedule under a single CWP and tally the total hours and duration it will probably add up to a different set of values than the CWP estimate. Then looking at this from a project management level you can decide if the longer (or shorter) duration is in line with the project objectives or if we would like to reduce (crash) the duration by adding more resources, run activities in parallel (fast track) or live with the longer schedule.

While the total sum of the IWPs in the level 5 schedule will help us to understand the burn rate that we will need to establish, the actual sequence of execution will probably change again based upon the reality in the field. This brings us to the Three Week Look Ahead (TWLA) and the pulse of the project. Eventually the Construction team stop looking at the level 5 schedule and think of it more as a pool of scope that feeds the hopper that is the TWLA, which is where the real story comes out.

Three Week Look Ahead								
	This week (In Progress)		Next week			Week two		Week three
OSBL	Sphere							
			IWP	Description				
	PE3-57-PI-10	Weld connections between Modules 3	PE3-57-PI-02	Install Large bore pipe on lower level o	PE3-57-PI-03	Install Large bore pipe on mid level of	PE3-57-PI-03	Install Large bore pipe on top level of A
	PE3-57-SS-01	Install Steel for lower level in Area 57	PE3-57-SS-01	Install Steel for mid level in Area 57	PE3-57-PI-11	Weld connections on LB LL Area 57	PE3-57-PI-12	Weld connections -2 on LB LL Area 57
	PE3-57-SC-25	Erect Scaffolds 36,38,39,40 &41	PE3-57-SS-02	Install ladders and platforms for lower	PE3-57-PI-14	Bolt up connections on LB LL Area 57	PE3-57-SS-02	Install light Steel for upper level in Are
	PE3-58-CO-15	Paving around DB36A	PE3-57-SC-25	Erect Scaffolds 37,42,44	PE3-57-SS-02	Install Steel for upper level in Area 57	PE3-57-SS-02	Install ladders and platforms Area 57
	PE3-58-CO-16	Foundations for CS03	PE3-58-CO-16	Foundations for CS03	PE3-57-SC-25	Erect Scaffold Tower for EB3 & Scaffol	PE3-57-SC-25	Erect Scaffolds 45,47,48 ,49
	PE3-58-EW-12	Grading Nth end of PR	PE3-58-CO-18	Paving sections 2, 3, 4 under PR Nth er	PE3-58-CO-16	Foundations for Pumps part A	PE3-58-EQ-31	Set Modules 124, 125, 126, 127
					PE3-58-CO-19	Paving sections 5 thru 9	PE3-58-CO-16	Foundations for Pumps part B
					PE3-58-SC-25	Hoarding for heat foundation Part A	PE3-58-CO-16	Pour and cure foundations part A
	Cooling Tower							
	PE3-59-PI-04	Dress SB Pipe on Stripper tower	PE3-59-EQ-29	Set Modules 124, 125, 126, 127	PE3-59-EQ-30	Set Modules EB Tower	PE3-59-SS-04	Install Ladders and Platforms East end
	PE3-59-EL-17	Terminate instrument cables on Stripp	PE3-59-SS-04	Install steel for expansion loop	PE3-59-SS-04	Install light steel in PR	PE3-59-PI-10	Weld connections between 12
	PE3-59-SS-04	Install Ladders and Platforms on stripp	PE3-59-SC-25	Erect scaffold for module interconnect	PE3-59-PI-10	Weld connections between 124,125	PE3-59-SC-25	Erect scaffolds 234 thru 241
	PE3-59-EQ-28	Set Modules 119,120,121, 122,123			PE3-59-SC-25	Erect scaffold for module interconnects		
ISBL	Pipe Rack							
	PE3-60-EW-16	Drive piles for Heater Structure	PE3-60-EW-22	Cut and Cap piles for heater structure	PE3-60-EW-18	Excavate for undergrounds NE corner	PE3-60-PI-04	Install and test Undergrounds NE
	PE3-60-EW-21	Cut and Cap piles for tanks	PE3-60-EW-17	Install duct bank	PE3-60-EL-17	Set up pull for underground cables	PE3-60-EL-17	Pull underground cables
			PE3-60-EL-17	Install duct bank	PE3-60-CO-15	Paving around NE PR	PE3-60-EW-23	Backfill and grade NE after hydro and c
			PE3-60-CO-15	Paving around DB36A	PE3-60-CO-16	Pour and cure Foundations for CS03		
			PE3-60-CO-16	Foundations for CS03	PE3-60-SC-25	Hoarding for heat foundation Part A		
	Fractionation							
	PE3-61-PI-03	Install Large bore pipe on mid level of	PE3-61-PI-03	Install Large bore pipe on mid level of	PE3-61-EW-17	Align pumps part A	PE3-61-EW-17	Align pumps part B
	PE3-61-PI-11	Weld connections on LB LL Area 61	PE3-61-PI-12	Weld connections -2 on LB LL Area 61	PE3-61-PI-03	Install Large bore pipe on mid level of	PE3-61-PI-03	Install Large bore pipe on mid level of
	PE3-61-PI-14	Bolt up connections on LB LL Area 61	PE3-61-SS-02	Install supports for field run tray	PE3-61-PI-12	Weld connections -2 on LB LL Area 61	PE3-61-PI-12	Weld connections -2 on LB LL Area 61
	PE3-61-SS-02	Install Steel for upper level in Area 61	PE3-61-SC-25	Hoarding for heat foundation Part A	PE3-61-SS-02	Install supports for field run tray	PE3-61-SS-02	Install supports for field run tray
	PE3-61-SC-25	Hoarding for heat foundation Part A			PE3-61-SC-25	Hoarding for heat foundation Part A	PE3-61-SC-25	Hoarding for heat foundation Part A

Trade	Colour Code	Ident
Scaffold	Orange	SC
Foundations	Black	CO
Structural Steel	Green	SS
Pipe	Red	PI
Mechanical	Yellow	ME
Equipment	Blue	EQ
Electricial & I	Grey	EL

The TWLA is created and maintained in a weekly development meeting where each Superintendent adds new IWPs to week three from the backlog, based upon their understanding of preceding work being completed and released to them. This triggers the Workface Planner to start working the secondary constraints: Scaffold, Equipment, Safety, Quality, Workforce and Permitting, which will all be addressed before the IWP is allowed to enter week one (the week before execution). The Workface Planner also notifies material management that the material that was reserved for this IWP can be bagged, tagged and staged.

TWLA format: We have gone backwards and forwards on the best format to use for this and we have settled on the idea that for the Superintendent when they are developing their ideal strategy it is best to use Excel. It is easy to apply and almost anybody can make it work. However, when all of the Superintendents have developed their individual nirvanas then the Construction management team need to herd these kittens into a single integrated three-week schedule with links and dependencies, and that only works in P6 or Microsoft Project.

When you do get to this point and you have an integrated TWLA that gets laid out in a Gantt chart from P6 and then played in a 4D simulation from the WFP software, you will be jettisoned into a new error of construction management. Be aware that once you have tasted this cool-aid nothing else will quench your thirst.

F. Field Level Execution

Then the day finally comes after all of these months and pages of preparation, the Superintendent gives the IWP to the Foreman (insert angelical music♪♫). There was a time when we thought that this was the finish line, but now we know that it is not. One of our key lessons learned was that it is not enough to give the foremen an IWP with the blind faith that it will be done in a week. These are construction professionals who <u>we</u> have trained to live hand to mouth in a system that makes them look like fools every time they try to plan and now we expect them to go from 0 to 60 in 3.9 seconds, (You might want to temper those expectations).

Getting the foremen to the place where they trust the IWP and can start to think about how to sequence activities over a period of one week rather than one day is a challenge. The best way that we found to address this was to apply a system that good superintendents already use: Daily 15-minute planning meetings at the start of each day before the crews arrive. We developed a simple Whiteboard format that lays out the plan for today and the actuals from yesterday. Each Foreman makes a projection for what they will achieve and what they will need, then they report the next day on how they did against the plan. It only takes a couple of weeks for the Foremen to get good at planning especially when their peers are watching.

Daily Foremans Plan - Whiteboard

				Mon			Tue			Wed			Thur			Fri		
		IWP#	Desc	Plan	Act	Equip	Plan	Act	Equip	Plan	Act	Equip	Plan	Act	Equip	Plan	Act	Equip
Civil	Crew 1	452C12	Foundations	50%			65%											
	Crew 2	452C13	Pedastals	15			12											
	Crew 3	452C14	Foundations	65%			70%											
Steel	Crew 4	431S01	PR Cloumns	12T		65T	18T		65T									
	Crew 5	431S02	Stairs C1	75%		35T	85%		35T									
	Crew 6	431S03	PR brace	28		15T	26		15T									
Pipe	Crew 7	652P04	Rigging	12 Spools		125T	8 spools		65T									
	Crew 8	652P05	Welding	32"		Manlift	18"											
	Crew 9	652P06	Welding	42"			20"											
Electrical	Crew 10	278T09	Tray	245'			390'											
	Crew 11	278T10	Supports	12			10											
Scaffold	Crew 12	1009S76	Tower	65%			100%											
	Crew 13	1009S77	Mods, A12	12			A15											
	Crew 14	1009S78	PR	4			7											

So that brings us to the delta between what was planned and what actually got done. The gap between these two points is either delays or incompetence and both can be addressed. We will spend some more time on delays in the next chapter on productivity, for now I would just like to mention that each foreman needs a story that explains the gap. The place to capture this is on the daily time sheet as a delay with an estimate for the hours of disruption that it created. These incidents and the weight of deviation can be mapped and reported so that the construction management team can understand the magnitude of the problems and address them.

If the gap has no tangible explanation, then the Superintendent needs to take a closer look and see if they need to support the Foremen with some coaching or training. Either way the system is very simple and works very well. It is also the epitome of the project management standard: If you track it, you manage it.

So now we reach the end of the week and the IWP is scratched up with notes and progress and the work is mostly complete. It is at this point where we need to apply the very important rule of the **"Use By"** date. This means that come hell or high water the IWP must be returned to the Workface Planner at the end of the cycle, whether the scope is complete or not. This is a very difficult rule to apply and we did end up losing this battle most of the time, but if you don't get the IWP back then the foremen end up with a library of them that are all at various stages of completion and nothing gets completely finished.

Incomplete plans that are returned to the Planner will tell you one of several things, either your plan was not good, the foreman choose not to follow it, your estimate for the required effort was wrong or something derailed the plan that the Planners did not anticipate. Whatever the case if the plan is not returned you will never know what the problem was, or what you could do to fix it.

We have had several projects where the closure of IWPs was mandated and the results were very good. Finishing the work in sequence and completely so that the crew does not have to return is a very productive way to execute scope. You can achieve this by having a progress hold back of 10% for each IWP that gets released when the Quality Control inspector has passed the work as compliant and complete. This works very well for shaping a culture of completions and will give you a very smooth transition into turnover. It also keeps the QC guys engaged and in the field.

What happens when stuff doesn't fit:

One of the tangible benefits of having the Workface Planners scrutinize the scope before it is released to the field is that you can anticipate that fully half of the problems will be addressed before they go to the field. We know this because we track the "Requests For Information" (RFI) and see that typically more than half of them are submitted by the Workface Planners. While this is very good news, it still means that almost half of the problems get discovered in the field, so you do still need a process that manages problems when they appear. We tried a range of different options and settled on the idea that if we have a problem that cannot be resolved on the same day then we create a new IWP. This allows us to treat it as a discrete scope of work that may require, engineering, materials, hours, planning and scheduling. At the end of the project, when the scope is complete and we want to know why there is a gap between what we planned and what we did, there will be a neat little pile of answers that are RFI IWPs. Quite literally worth their weight in gold.

The last step in the execution process is that the Workface Planner takes the returned IWP and enters the progress into the WFP software, removing any left-over scope and putting it into a cleanup package.

G. Tool Time studies: Addressed in detail in the next chapter: Productivity Measurement

Module Construction

Just before we leave Workface Planning I wanted to point out some common sense that is not common practice around the productivity in module construction yards.

It matters and we can do something about it.

The logic that it is not part of the construction project and is therefore exempt from being optimized is simply not true and anybody who has suffered from the late or out of sequence delivery of modules will tell you the same. 'Somebody should do something about that', well that somebody is us and if you apply this chapter to the mod yard, the critical path (Pipe Rack) of your project will dramatically improve.

While you are there make sure that everybody understands that the reason that we build modules off site is because it is cheaper, faster and safer than site construction, so don't ever, ever, ever allow anything that could be done in the yard to get transferred to site.

It is OK to optimize the yard, but not at the expense of site construction.

Chapter 8: Productivity Management

Productivity is never an accident. It is always the result of a commitment to excellence, intelligent planning and focused effort. Paul J Meyer.

It may be a shock to hear me say this but, the total management of productivity is bigger than just AWP. A project environment that is based upon the sequential execution of work packages will create an absence of chaos, which is a good place to start. The development of this stable workplace will then give rise to other opportunities and you may find that people want to step up to take advantage of that.

A good example of this is the material management processes on one of our previous projects, where the yard manager took our plan to the steel fabricator and loaded his trailers by IWP. The trailers where then shipped to the yard and left on the trailer until we called for the material. The yard manager understood that the cost of trailers was very small when compared to the cost of labor to unload and reload the materials. In this situation, the creation of IWPs prior to the steel fabrication allowed him to take advantage of a predictable environment and reduce his need to double handle the material. While this type of optimization is not the primary target of AWP, it is the extended value that your entrepreneurs will take advantage of when you give them the opportunity.

Our experience has shown us that the creation of a stable construction environment will give rise to many more opportunities like this one, however you may also find that your entrepreneurs are slumbering under the weight of traditional construction and they may need a bit of a push to help them get started again. That push can come in the form of productivity measurement, which alerts people to the idea that we may not yet be at peak performance and that there is probably room for innovation and incremental improvements, outside the work packaging effort.

In the following pages, we will explore several ways to track and influence productivity and also have a look at some of the environmental factors that can contribute to an aggregation of marginal gains.

Sections

A. Contracts
B. Owner Engagement
C. Productivity Champion
D. Tool Time Analysis
E. Delay Codes
F. Earned Value Management
G. Key Performance Indicators

A. Contracts

Let's start with the idea that productivity can be influenced by contracts. The logic is that a good contract establishes incentives for behaviors that lead to good productivity. Lump-sum/unit-rate contracts are a good example of this where the risk and reward are transferred to the contractor so that they can deal with it. It looks like the 'Easy Button' for Owners. The trouble with this is that it assumes that the contractor has control over the things that influence productivity. The contractors know that they don't so they add contingency to the price (which is the risk that the owner thought that they wouldn't have to pay for) and then make claims for every impediment caused by factors beyond their control. The owner pays the inflated lump-sum price and all of the claims. And worse than that is that the contract didn't drive the productive behaviors that were targeted, the contractor's focus was on capturing claims not on being productive.

The conclusion of this scenario is that there is no such thing as the 'Easy Button' for Owners and contract incentives that you thought would drive productivity typically encourage the wrong behaviors.

My own view of the world is that lump-sum contracts only work for things that are small, definitive and common. The construction of tanks and cooling towers are good examples, less than $50 Million, off the shelf design and the companies that do these, do them all the time. While this is no guarantee of success, the probability of a predictable outcome is quite good, if the supply lines are wide open. Having said that, this is still not the 'Easy Button' for project management and if there is going to be a positive result it will still require active engagement from the Owner led project management team to ensure an uninterrupted flow of information, materials, tools and access.

Considering the idea that lump-sum contracts are used when there is a definitive path that allows the contractor to execute scope unimpeded and without consideration for other parts of the project, this style of contracting certainly doesn't work for engineering, procurement or construction management. In an effective project environment, these disciplines must all be 'in service' to construction, which means that if the contract type drives behaviors that encourage them to optimize their own systems it will be at the expense of construction, which is the exact opposite of what they are supposed to be doing. The industry has recognized this and time and material contracts are almost always used for these disciplines.

The aim of a contract is to establish an expectation on both sides for an outcome that is mutually beneficial to both parties, win-win, to quote Stephen Covey. The extension of this logic is that any contract that is established to present a win-lose outcome (Where the contractors assumes all the risk) will end up as a lose-lose, which typically ends up in court where the only winners are the lawyers.

So, the message is that while lump-sum or unit-rate contracts have their place in some pockets of the project, by themselves they are not enough to tackle productivity.

At the other end of the spectrum we have Time and Material contracts that on the surface look like we have given up on the idea that we can drive productivity through contracts. However, T & M contracts can still be developed with requirements and incentives that drive the desired behaviors.

We need to understand exactly what it is that E, P, CM and C do or don't do that influences field productivity and we need to develop contracts that reward those positive behaviors.

Engineering's influence on construction productivity is probably a mixture between producing good quality engineering and doing it at a rate that stays ahead of construction = error rate and schedule compliance.

We do know that perfect engineering is far too slow and expensive, and that engineering errors are very disruptive to cost and schedule. So, the contract should strike a balance between errors and the on-time delivery of EWPs. The sweet spot between these points is probably an error rate of < 2% with >95% schedule compliance on complete EWP deliveries. Of course, this also means that you must track this and be realistic with your EWP due dates.

For Procurement, the criteria required to support construction productivity is probably very similar where there must be an acceptable level of quality control that captures more than 98% of issues while also meeting schedule dates for complete PWPs at least 95% of the time.

If these conditions were mandated through contract terms then we could envision a situation where before they start work on any single CWP, the construction contractor would have more than 95% of the materials that have less than 2% engineering errors and 2% fabrication errors. This would lead us to believe that there would be a very favorable environment to execute construction.

The execution of work will still be influenced by scaffolds, cranes, manpower, permits and other owner issues, so we are not in the promised land that would allow lump-sum to work, but we would have a situation where the two biggest productivity impediments had been substantially addressed.

It is easy enough to identify the areas of influence that the Construction Management team have on field productivity: the supply of CWPs, cranes, scaffold, site services, material management etc. But it is far more difficult to develop a set of metrics that show how effectively these services are being provided. There are not many examples of CM contracts that drive the right behaviors, although it seems like something that was tied to the construction contractor's performance would encourage the CM team to facilitate contractor productivity. (See Tool Time studies).

The key is to shape contracts that target holistic project outcomes by considering the impact on construction (the motherload of project cost), not just the reduction of cost and schedule for each silo.

Incentives for the construction contractor need to focus on things that they have direct influence over: % of scope that is packaged in IWPs, compliance with the three week look ahead, rates of placement for each commodity, weld inches per welder, rework, total workforce, safety incidents, quality compliance and the number of punch list items. A T&M contract that was set up to reward achievable targets in each of these areas would drive the contractor to prepare for and execute work in the most optimal format.

In summary and from the perspective of productivity, contracts can be used to develop a foundation that encourages the right behaviors but they are still not the course correction that will solve world hunger. The key is to develop contracts that target behaviors that support productive construction, even if on the surface it looks like a cost increase. This can set the foundation that will allow the Owner led project management team to be effective.

B. Owner Engagement

The next area that I would like to explore for marginal gain is Owner engagement. Interestingly amongst the companies who track project performance, good quality planning and Owner engagement have been identified as the top two influences on project success.

Each Owner company seems to cycle through their level of engagement with their project teams, starting out with the idea that minimal involvement allows the contractors to do their thing, which is why they were hired. Then after the perceived fleecing and frustration of near zero influence, the Owners go to the other end of the scale on the next project and try to do the whole thing by themselves, which typically disproves the idea that the first one couldn't get any worse. Armed with these real-life case studies the Owners (If they are still in business) develop a blended team that taps the expertise and process systems of the contractors and mixes it with a sprinkling of Owner imbeds who occupy strategic positions. Sometimes this starts out as a duplication where each contractor department head has an Owner counterpart, this creates an adversarial relationship that brings some balance and scrutiny to decision making, but still pits us against them.

The most effective models are where the owner is imbedded into an integrated team, where the contractor supplies process, people and expertise and the Owner supplies the guidance and holistic vision for project success.

We have personal experienced with this model on several projects and it works really well for us. Typically, we represent the Owner and in our case, we supply the process (because AWP expertise is still developing amongst the contractors).

The integrated teams that we have seen function in other departments have similar results which stem from team decisions and actions that are founded upon years of experience, targeting desirable outcomes for the whole project.

This does mean that the contractor must allow the Owner to play in their sandbox, but it also, almost always guarantees a certain level of success. It is very difficult for the Owner to blame the contractor for results that they had influence over.

One of the other major contributions that the Owner representative brings to the team, is that they provide a voice with the rest of the organisation and that their connection to the project management team can be used to drive issues that are important to their department.

The development of the integrated team does take some design and coaching, if the Owner representative is too autocratic or too laissez faire, then the team direction does not have the balance that it should have. At the end of the day the team is still not a democracy and as the appointed lead the Owner representative has the responsibility to make decisions, but the smart ones will do so with the support of their contractor team.

To operate effectively in this position the Owner representative does also need to have some experience in this field, so that they can offer guidance and be guided.

Our experience with Integrated teams has been that for them to be effective they must also be co-located and even, if it is possible, have the same email address: John.Hancock@project.grd

We were on a project were the integrated team went so far as to ban other company clothing, but that also meant that they had to supply project jackets and shirts.

To prove the point of integration being effective, think back to a project where an Owner representative commented after the project, that their contractors worked well with them. This probably means that they took advise from the Owner and that they also offered their own advice based on their experience and that this cooperation led to mutual gain results.

C. Productivity Champion

Just before we move onto some more traditional ways to address productivity there is another key environmental factor that we should discuss: The unhindered flow of Tools, Materials and Access for the contractors. This subject circles back to contracts and needs to be one of the considerations when the Project Management team are establishing an overall contracting strategy.

A simple example is tools: We know that a worker needs tools to execute work, we also know that the average cordless drill cost around $200 and that the total burdened cost of labor in some of our markets is also around $200 an hour.

If we ask the T&M contractors to supply their own tools for a markup on each hour charged, then the contractor can increase their profit by minimizing the amount spent on tools. If there is a shortage of tools in a T&M contract then the delays are paid for by the owner, through a lack of production. So, a contracting strategy (get the contractor to supply their own tools) that was aimed at reducing Owner costs, is actually promoting the exact opposite outcome.

A much better train of thought is to understand that tools are cheap and that labor is expensive, so we should have 3rd party tool vendors onsite supplying everybody with whatever it is that they ask for. The tool vendors have simple tracking software that tracks tools against each worker by scanning their badge so any abuse can be isolated and addressed. The small amount of abuse that is not addressed is inconsequential in the grand scheme of labor costs. The incentive for the 3rd party tool vendor is to supply as many tools as they can, which leads to every worker having all the tools that they need: problem solved.

The same is true of bolts, nuts, gaskets, electrical widgets, gloves, glasses and even hard hats. The cost of each is insignificant compared to the cost of labor and all of these items can be easily stocked and distributed onsite by 3Rd party vendors. The Owner removes the allowance for tools and consumables from the hourly rates that they pay to the contractors and pays the vendors directly. It is in the Vendor's best interest to keep their trailers stocked, staffed and to track all transactions. The project ends up with a steady supply of the right tools in the right hands without any significant delays. This is a good example of a win-win situation where the Owner engages specialty contractors to ensure the supply lines are kept wide open to the workers.

Now let's apply the same logic to scaffolds, cranes, aerial work platforms, welders, temporary power and even permits. These are all components that populate the list of worker requirements needed to execute scope and they are also traditionally the things that are an obstacle to construction productivity because we try to minimize their cost.

Accounting 101:

If a segment of scope needs a scaffold, crane, power supply, permit etc. before it can start then it is already a sunken cost and the only control that we have over it is to try to supply it fast enough that it does not also incur a variable cost, which is the cost of a delay. Therefore, restricting the supply of any of these commodities only delays the cost and then exacerbates the problem with the additional cost of poor productivity and claims.

Unfortunately, there is no silver bullet solution for these issues, but we have seen them all successfully addressed individually. The real key to the solution for these and a host of other, non-AWP issues, is to have a Productivity Champion. Somebody who understands how systems work, who can analyze inputs, outputs and the cause and effect cycles behind undesirable results.

Which leads us to the question: How do we identify undesirable results? To answer that question, I am going back to a quote from Yogi Berra:

"You *can see a lot just by looking"*

Translated into construction speak this is Tool Time Analysis.

D. Tool Time Analysis

Time on Tools studies are a systemic process that captures a numerical snapshot of the activity levels of the trade workers on a construction project. Typically conducted every two months during the construction phase they show actual results and trends that can be linked to project events.

A trained observer (The Productivity Champion) makes several thousand random observations of the entire workforce over a two-week period, categorizing each observation as being engaged in:

Direct: Activities that lead the project towards completion
Support: Typically Planning and Travel that support the preparation to execute direct work.
Delay: Worker is delayed due to the absence of Information, Materials, Tools, Access or Desire.

Sub categories break down the observations into smaller groups for further analysis:

Support
1. Travel
2. Planning

Delays
1. External: Worker is delayed by another trade or company
2. Internal: Worker is delayed by their own team
3. Personal: Worker is delayed by their own choice
4. Supervision: Worker is delayed by their own supervision
5. Equipment: Worker is delayed by an absence of equipment or tools
6. Materials: Worker is delayed due to an absence of material

Importantly the process has been utilized on construction projects globally by a wide variety of organizations over the last 30 years so it is recognized as the universal platform for industrial engineering and productivity analysis.

The other really important facet of the results is that they are not linked to the estimate, site factors or the subjective nature of progress reports. They are a black and white snapshot in time that records actual engagement or not.

The extended value is that the observer will try to capture the root cause of the delays, either by recording the obvious or by talking to the workers. By the end of the survey this rolls up to a series of cause and effect features that the observer reports with the chart numbers.

The individual results are recorded in a database, which allows them to be sorted by contractor, trade, area, day or time of day. This leads to specific analysis of unique circumstances and because the figures are proportional to a day's work, the true value of those specific influences.

As an example, if one area has a high incidence of travel and the observer noted that the lunch room facilities are placed further away from the workface than other instances on the project, we could calculate the differential as a portion of the day and extrapolate the cost of the extra travel time.

Now imagine the same scrutiny applied to the supply of information, materials, tools, scaffolds, cranes, aerial work platforms, welders, temporary power and even permits. Typically, these common factors are captured during the studies and reported along with the overall results.

The power of the Tool Time Analysis is that it isolates these cause and effect cycles and shows their impact on the project, which allows the Construction Management team to focus on high value issues.

The Productivity Champion is then often sent back to develop a comprehensive study of the specific systems that map the processes and identify opportunities for improvement.

I mentioned earlier that it is traditionally difficult to develop metrics that record the effectiveness of Construction Management, because of the wide variety of influences. However, the levels of worker activity recorded in a tool time analysis are essentially a score card for the effectiveness of Construction Management. This is based upon the proven premise that workers will do as much as we allow them to do.

Our proof for this is any typical welder:

In a fabrication shop where the welder has a controlled environment we have a standard expectation that a single welder will weld 100 diameter inches in a day. On an industrial project, we are very excited when we can get the welders to achieve 10 diameter inches per day. The welder is the same person, with the same capacity, the only thing that changed was the controlled environment. The welder onsite cannot control how efficiently they receive drawings, materials, scaffolds, temporary power or permits, these are all controlled by the construction management team. Knowing this, it is easy to understand that while the Tool Time Analysis is a record of worker activity, the results are owned by the Construction Management team.

The relationship could be compared to a sports team, where the coach is ultimately responsible for the result. They achieve success by optimizing the players and bringing harmony and balance that produces results.

Maybe sometime in the future we will be able to tie a contract incentive program for Construction Management to the results of the Tool Time Analysis.

In the meantime, on the projects where we have applied both AWP and Tool Time Analysis, we have seen specific behavior changes that target the issues that are identified in the studies. Especially when the report is shared with the entire project, so that construction management peers can compare their results.

The data that comes from tool time studies also helps us to understand the value of small incremental changes and the true aggregation of marginal gains. Take for instance the difference between having two 30-minute breaks on a construction site and the traditional two 15-minute and one 30-minute breaks. Mapped over a 10-hour day (on the following page) you can see that the difference appears to be an insignificant 20 minutes, but when filtered by the view that only 40% of the day is spent in direct work, an extra 20 minutes at the workface is a 2% increase in tool time.

To put this into perspective, a rule of thumb for the total cost of the Workface Planning portion of AWP (Planners, software, hardware and training) is that it is equivalent to 2% of the cost of construction, which is a little more than 1% of the Direct Activity from a Tool Time study. So, if you wanted to cover the cost of WFP, you could apply the two-break policy to your construction workforce and you will have doubled your investment, even before you start to reap the benefits of the returns that will come from work packaging and information management.

Apart from the mathematical advantage the two-break system is also much preferred by the workforce, so this is a good place to start your journey towards optimization.

Some of the other data that comes out of the Tool Time studies is that we have higher productivity in the mornings and in periods where there is a continuous period of work. Hence the 200 minutes of continuous work in the morning of the 'two-break' schedule. Another way that one of our projects used this data was to recognize that getting started and being uninterrupted are important so there was a total ban on any meetings for field supervisors between 7 and 11am.

As a leading indicator, the Tool Time studies also allow the Construction Management team to address actual current issues and to see the results of their actions in a much shorter period of time (typically productivity changes take up to a month to show up in earned value management reports).

2 x 15 min breaks and 30 mins for lunch.

	Activity	Time at the workface		
6:30	Foreman daily meeting			
7am	Boots on and ready for work			
7:30	Start work meeting complete, tools in hand			
	Work period	110		
9:20	Tools down, travel to facilities for break			
9:30	Coffee break			
9:45	Return to workface			
9:55	Tools in hand			
	Work Period	115		
11;50	Tools down, travel to facilities for break			
Noon	Lunch break			
12:30	Return to workface			
12:40	Tools in hand			
	Work period	130		
14:50	Tools down, travel to facilities for break			
15:00	Coffee break			
15:15	Return to workface			
15:25	Tools in hand			
	Work Period	100		
17:05	Tools down, travel to facilities for clean up			
17:15	Clean up period		Direct	
17:30	End of shift		activity	
	Total time at the workface	455	x 40%	182
	Out of	600		600
		76%		**30%**

2 x 30 minute breaks

	Activity	Time at the workface		
6:30	Foreman daily meeting			
7am	Boots on and ready for work			
7:30	Start work meeting complete, tools in hand			
	Work period	200		
10:50	Tools down, travel to facilities for break			
11:00	First Break			
11:30	Return to workface			
11:40	Tools in hand			
	Work Period	130		
13:50	Tools down, travel to facilities for break			
14:00	Second Break			
14:30	Return to workface			
14:40	Tools in hand			
	Work period	145		
17:05	Tools down, travel to facilities for clean up			
17:15	Clean up period		Direct	
17:30	End of shift		activity	
	Total time at the workface	475	x 40%	190
	Out of	600		600
		79%		**32%**

E. Delay Codes

If you track it you manage it.

The idea of tracking delays is a simple one that is applied by all of the good contractors. Normally they do it so that they have a story when the Owner asks them about the gap between what they scheduled and what they achieved. However, if tracked and recorded diligently, they also know that it leads nicely into claims and support for budget and schedule discrepancies.

The real problem is that while all of this tracking has been going on the construction management team usually don't address the issues. The logging of delays is viewed as contractor whining that is largely unsubstantiated and probably exaggerated, so it is ignored.

On the few projects where we have been able to mandate delay tracking and reporting on the timesheets, we end up with a colossal amount of almost quantified data, that points straight at unresolved construction management issues.

There is an obvious reluctance from the Owner to endorse the number of hours reported lost by the contractor to permit delays or the absence of scaffolds. However, the true value in the report is in understanding that there is an issue and that the hours reported show the magnitude of the problem.

The execution is quite simple: Contract language is added to all contracts requiring the Foremen to record any delays on their timesheet by total lost hours, based on a standard matrix which is printed on the back of every timesheet. The contractor records the delays and reports them each week as an element of the weekly report.

And then the construction management team do something about it.

If you track it you manage it. (and the opposite is also true)

DELAY CODES

Delay Title	Code	Description
Safety - Concerns	**S 1**	Task could not be performed safely
Safety - Inadequate Training	**S 2**	Safety training requirements not met.
Plan - Preparation Not Complete	**PL 1**	Preparation for task execution inadequate.
Plan - Preparation - Scaffolding	**PL 2**	Delayed by the lack of scaffold or modifications
Plan - Trade Coordination	**PL 3**	Access to the workface restricted by other trades
Plan - Work Scope Insufficient	**PL 4**	Insufficient instruction, scope not identified / understood.
Plan - Change to Workfront	**PL 5**	Workforce delayed due to unplanned change in workfront
Rework - Fabrication / Engineering	**RW 1**	Components do not fit
Rework - Workmanship	**RW 2**	Rework task due to poor workmanship.
Resources - Material Unavailable	**RS 1**	Materials for the task were not available
Resources - Equipment Unavailable	**RS 2**	Equipment not available. (Cranes, lifts, welders, pumps)
Resources - Tools Unavailable	**RS 3**	Tools not available. (equipment under $1500 in value)
Resources - Trades Absent	**RS 4**	Workforce did not show up (sick/absent/late/).
Resources - Trades Unavailable	**RS 5**	Workforce shortages
Permit Delays - Issuer	**PM 1**	Permiting not efficient. (permit not ready at start of shift)
Permit Delays - Operating Unit	**PM 2**	Operating unit withheld permits for operational activities
Permit Delays - Other	**PM 3**	Permits not requested in time. (previous day)
Permit Delays - Unit Upset	**PM 4**	Permits cancelled due to Unit upset.
Travel Delay	**T 1**	Travel between facilities and the workface > 5 minutes.
Travel Delay - Vehicle	**T 2**	Travel delays created by lack of access to vehicles
Weather - Precipitation	**W 1**	Rain, Fog, Snow.
Weather - Wind	**W 2**	Wind creates a hazardous environment
Weather - Temperature	**W 3**	Too hot or too cold
Other	**O**	Explain

F. Earned Value Management

So how do we traditionally measure productivity?

In its simplest form, the measurement of productivity is Effort over Output. How many hours work did we spend and how much did we get done. In the first book, Schedule for Sale, I compared the Productivity of Canada to Uganda and how in 2008 both countries had about the same population, who contributed about the same amount of estimated labor hours but that Canada had about 100 times more output than Uganda. This was to demonstrate the true value of technology as opposed to just working hard. The Output, reported as Gross Domestic Product is a sum of the wheat, oil, cars and other stuff, consumed and exported. The Effort is the sum of estimated labor hours that it took to produce these goods.

In construction, the equation is the same with different units of measure. For Output, we use Barrels of oil or square feet of building space at a macro level and tons of dirt or meters of pipe at the field level. The effort is the amount of direct labor hours incurred by the workforce.

We also know that tracking productivity is very important to the industry, projects and the individual contractors as a common means of measuring effectiveness

Methods:

There is the Cost Performance Index (CPI) more commonly known as the Productivity Factor (PF), which measures earned hours against burned hours: $PF - \frac{\text{Earned}}{Burned}$. For most of the world this means that a factor above 1 is good. Computed at the right level (CWP), this gives you a very good idea of how the project is performing against the estimate, but it is still subject to the quality of the estimate.

If the estimate was developed based on a WAG then measuring performance against it, is only an effort to prove that the guess was lucky. If the estimate was based on actual quantities multiplied by standard installation rates then the measurement of performance could be meaningful, however this can also be misleading if the company's standard rates are different than the next company, which typically is the case. This means that companies are performing against their own standards for installation which may or may not be competitive with the rest of the market.

Installation Rates

		Carbon Steel			Stainless Steel		
		Std	XS	XXS	Std	XS	XXS
Metric	**Imperial**	**Sch 40**	**Sch 80**	**Sch 160**	**Sch 40**	**Sch 80**	**Sch 160**
15	**0.5**	0.2	0.2	0.3	0.2	0.3	0.3
20	**0.75**	0.3	0.4	0.4	0.4	0.4	0.4
25	**1**	0.5	0.5	0.5	0.5	0.5	0.6
40	**1.5**	0.7	0.7	0.8	0.8	0.8	0.9
50	**2**	0.9	1.0	1.0	1.0	1.1	1.1
80	**3**	1.4	1.5	1.6	1.5	1.6	1.7
100	**4**	1.8	2.0	2.1	2.0	2.2	2.3
150	**6**	2.8	3.0	3.2	3.1	3.3	3.5
200	**8**	3.1	3.9	4.2	3.4	4.3	4.6
250	**10**	3.3	3.5	3.7	3.6	3.9	4.1
300	**12**	3.5	3.7	3.9	3.9	4.1	4.3
350	**14**	3.6	3.8	4.0	4.0	4.2	4.4
400	**16**	3.8	4.0	4.2	4.2	4.4	4.6
450	**18**	3.9	4.1	4.3	4.3	4.5	4.7
500	**20**	4.0	4.2	4.4	4.4	4.6	4.8
600	**24**	4.1	4.3	4.5	4.5	4.7	5.0
750	**30**	4.5	4.7	4.9	5.0	5.2	5.4
900	**36**	4.7	4.9	5.1	5.2	5.4	5.6

Sample Data

X

Rules of Progress

	Receive	Install	Connect	Support	Test	Punch
Pipe	5%	20%	40%	15%	10%	10%

Daily Timesheet

Foreman	Bob Fitswell					
	Date		Thursday	June 12 2017		
WTF-I-12	IWP	IWP	IWP	Delay code	Delay code	Delay code
Name	P3-C02-11	P3-C02-12		W5		
Joe Conners	10					
Tom Wright	5	5				
Abdul Eladin	5	5				
Megan Moore	10					
Lui Chu	3	7				
Francesco Lotti	3	7				
Bubba Watson		10				
Jose Rodrigus		10				
Gregory Tightenof	2	5		3		
Juan Diaz	2	5		3		
	40	54		6		

= Burned Hours per IWP

= Earned Hours per IWP

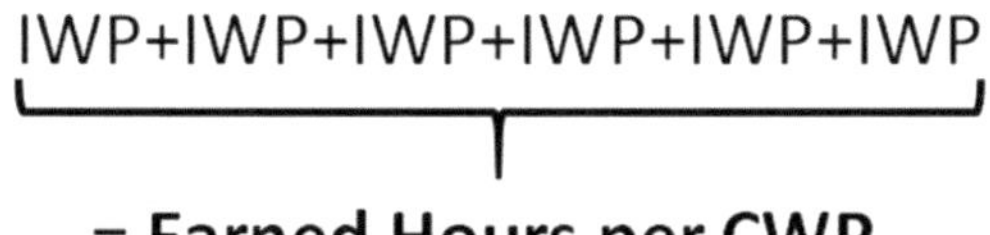

= Earned Hours per CWP

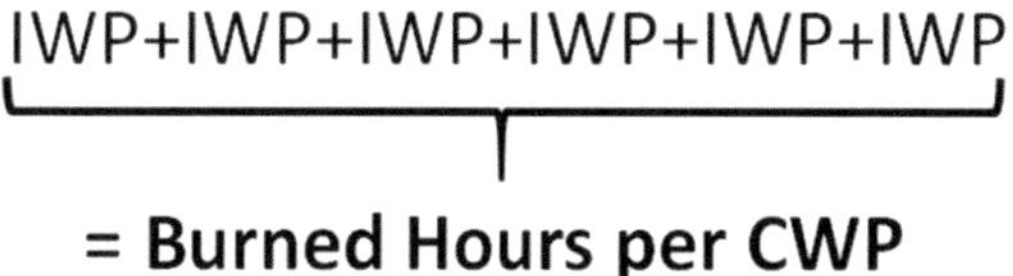

= Burned Hours per CWP

The same is true for performance against the schedule. You can track your schedule performance index (SPI) which shows if you are ahead or behind, but this is still based on the estimates for durations that may or may not have been calculated from quantities multiplied by installation rates, divided by resources. (The same logic that calculates it should take 9 women 1 month to have a baby).

This means that in the grand picture of project success, productivity factors like the CPI or SPI help to understand performance against the estimate but are not good indicators of our construction productivity. If we predicted that our productivity would suck and it did, then we achieve the estimate and we are on time and on budget.

Both of these systems for measurement (CPI and SPI) are useful in showing trends, if you don't mind a bit of lag. Mathematically it makes sense that if your timesheets and progress are reported at the end of each week then the calculation of performance should show how things went in the previous week, but it never seems to be that responsive. It normally takes an extra week or two to show a positive or negative trend, which is probably attributable to the human factor. Habitually field supervisors sandbag their progress so that they always have some in reserve for bad weeks, I know that I did when I was in the field. I didn't want to be that (Honest) person whose progress was always up and down.

Another way to track productivity is to use this same process of tracking quantities installed against hours burned, but to do so using industry standards rather than inhouse installation rates, as the benchmark for performance.

As an industry, we know that we can install concrete at 15 hours per cubic yard, steel at 35 hours per ton, pipe at 3 hours per foot and cable at .25 hours per foot. While each project is unique and there is typically a factor applied to these rates for complexity caused by height off the ground, congestion or extremes of weather, going forward this is our only true gauge of productivity that compares project to project across the industry. It calculates effort against output on a standard scale. Which means that our measure of success will be our proximity to the industry benchmark and not whether we knew how bad we would be or not.

The contractor's argument against using industry standards is that companies have their own standards which represent their competitive advantage and they don't like to share their numbers. Which doesn't make any sense to me at all, but the industry seems to accept this argument. It means that if a company routinely underperforms the rest of the industry then they don't want to be compared to the industry standards because they would lose any chance that they could get work.

Competitive advantage or smoke and mirrors?

Anyway that you look at it, this current disconnection between performance and success leads to a situation on many projects where it is more important to set a healthy budget with lots of schedule contingency, than to build a foundation for high productivity. It is a fundamental flaw in our efforts to become accountable and predictable, and something that we must address as an industry if we are going to be taken seriously by the rest of the world.

The third facet of overall effectiveness is **cost**. The logic applied in markets where the wage rates are much lower than the western world is that productivity is not as important as overall cost, which is somewhat true. Does it really matter if we need 4 times more people to execute a task if their wage rate is only 20% of the high productivity workers? Maybe not today, but the world is changing. Asian markets are experiencing a tidal change where the need for higher quality is driving costs up, just as it did in the western world over the last 100 years.

The balance between cost and quality is a conundrum that we all face. Is it better to hire a cheap plumber (if there is such a thing) and live with the risk that the basement might flood while you are on holidays.

The answer lies in our tolerance for risk, or more importantly the project's tolerance for risk, demonstrated by the fact that typically nuclear plants don't use cheap materials.

The reality is that productive techniques can already compete with low wage rates, that is why Canada is so much more productive than Uganda. I also believe that we will double or triple our productivity in construction over the next ten years as we optimize our use of technology and AWP, while the wages and quality of work in low wage markets will steadily increase, closing any gap that is there now. So if you are banking on the idea that low wages and low quality will be able to compete with high productivity and high quality, your gamble is a short term, short sighted, risky strategy.

Which means that the 'cheap labor = don't worry about productivity' argument is shallow and has other elements that can erode any cost savings.

So, lets continue to worry about productivity.

G. Key Performance Indicators (KPIs)

What do you need to know to understand productivity on a construction project? This is a question that we posed to a collection of 25 project managers after a $10 Billion project in Canada in 2006. Specifically looking at productivity and excluding safety, quality, engineering and procurement, because their reporting functions are more refined. We ended up with an extensive list of elements and then spent the next several projects boiling it down to a standard stewardship template for the weekly contractor meetings. For our new projects, we start with this model and then adapt it to suit the unique characteristics of each project.

The following list of our standard KPIs and methods for communicating them is a view of the finished report. Behind each graph there needs to be a structured environment that generates the stable data, which in our world is mostly satisfied by the process of Workface Planning and the creation of IWPs in the WFP software.

Total Workforce: Total number and breakdown of trades people by week.

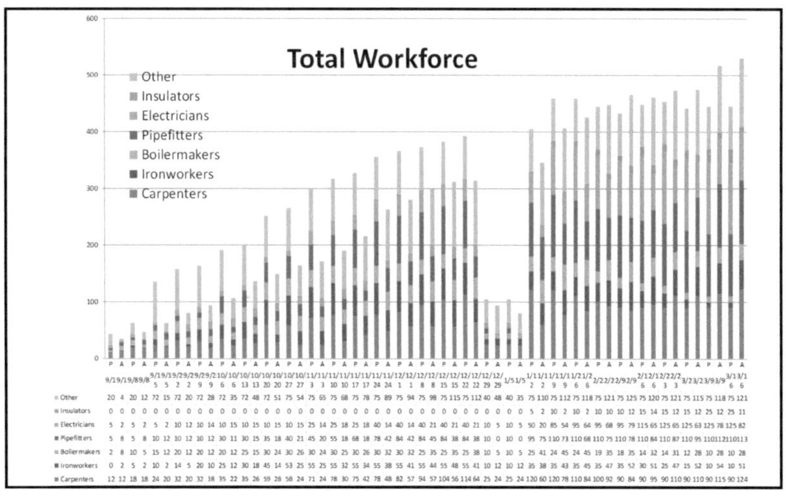

If you want to execute scope, workforce is the first requirement. This can be a major constraint in a tight labor market or in a market where there is a high percentage of unskilled workers.

Can also be used to show the average ratio of Supervisors and apprentices/helpers.

Overall Productivity: Earned over Burned = PF (CPI) by week and cumulative, shown against a backdrop of planned and actual overall % of progress.

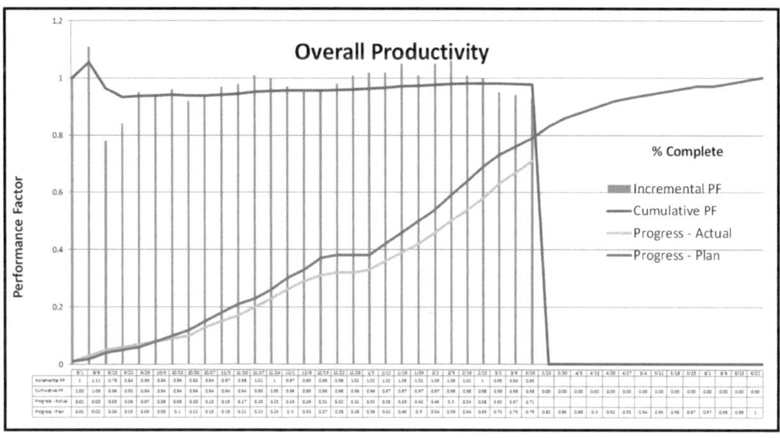

As described earlier the proximity of the PF to 1 is only an indication of whether the estimate was lucky or calculated. The real news is the cumulative trend. Trending up, keep doing whatever you are doing, trending down, have a hard look at your field execution work processes.

IWP Production: Tracks the rate of development of IWPs. Drawing data from Pack Track.

The AWP model calls for IWPs to be developed in a rolling wave schedule, where their development is no more than 90 days ahead of execution. This report shows their flow through rate and the backlog, highlighting areas for attention.

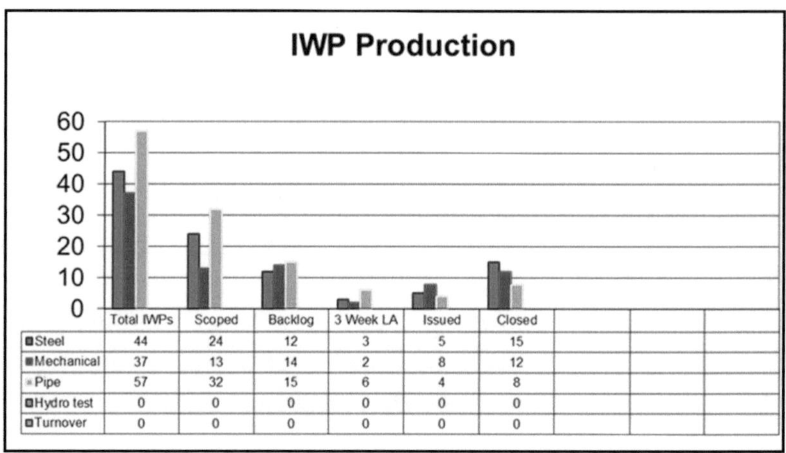

Delays: Reported by the foremen on the daily timesheets and summarized each week, these figures quantify the magnitude of common issues.

The contractors can also add issues to this report, (the story behind the numbers) listing specific issues that they feel are obstacles to productivity.

It's the Construction Management Team's 'to do' list.

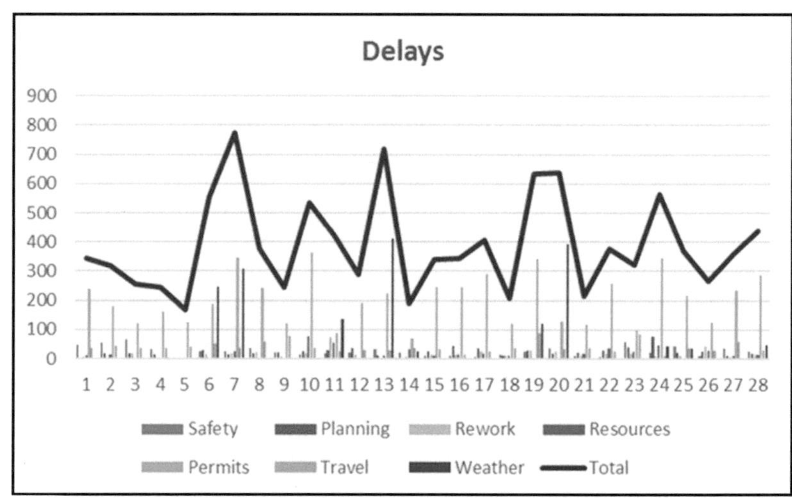

Indirects: The tracking of Indirects as a % of Directs is useful information, but be warned that on an AWP project the numbers are much different, because we have more planners and less workers, which is a good thing. Also shows the ratio of Scaffolders against Directs which shows how well the resource is being utilized.

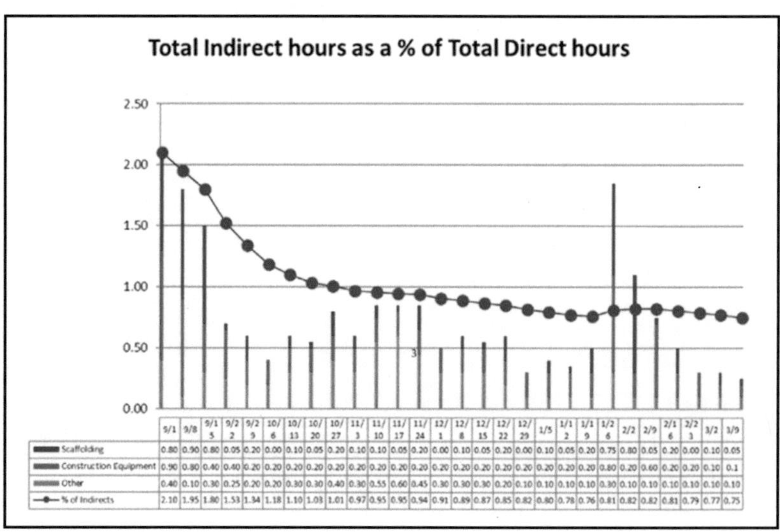

Scaffold: Tracks the backlog of scaffold by volume and the earn rate by volume, which tells you if you have enough scaffolders.

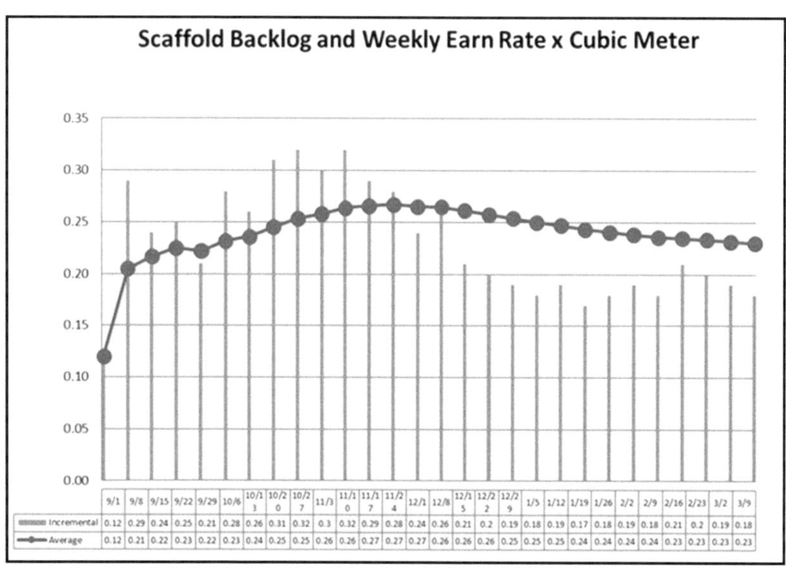

	9/1	9/8	9/15	9/22	9/29	10/6	10/13	10/20	10/27	11/3	11/10	11/17	11/24	12/1	12/8	12/15	12/22	12/29	1/5	1/12	1/19	1/26	2/2	2/9	2/16	2/23	3/2	3/9
Incremental	0.12	0.29	0.24	0.25	0.21	0.28	0.26	0.31	0.32	0.3	0.32	0.29	0.28	0.24	0.26	0.21	0.2	0.19	0.18	0.19	0.17	0.18	0.19	0.18	0.21	0.2	0.19	0.18
Average	0.12	0.21	0.22	0.23	0.22	0.23	0.24	0.25	0.25	0.26	0.26	0.27	0.27	0.27	0.26	0.26	0.26	0.25	0.25	0.25	0.24	0.24	0.24	0.24	0.23	0.23	0.23	0.23

Requests For Information (RFIs): Tracks the number of open RFIs, the average cycle time for responses and the total number.

The RFI process can be debilitating if it is not managed. This report typically becomes the center of attention at the weekly RFI review.

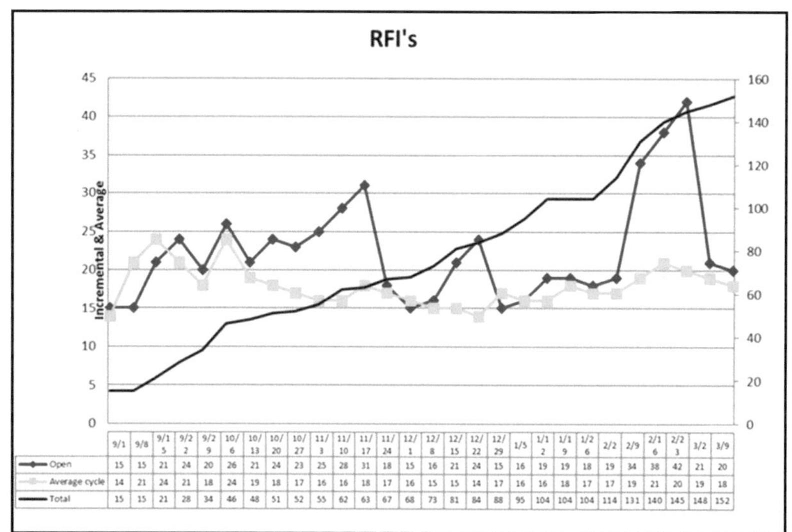

	9/1	9/8	9/15	9/22	9/29	10/6	10/13	10/20	10/27	11/3	11/10	11/17	11/24	12/1	12/8	12/15	12/22	12/29	1/5	1/12	1/19	1/26	2/2	2/9	2/16	2/23	3/2	3/9
Open	15	15	21	24	20	26	21	24	23	25	28	31	18	15	16	21	24	15	16	19	19	18	19	34	38	42	21	20
Average cycle	14	21	24	21	18	24	19	18	17	16	16	18	17	16	15	15	14	17	16	16	18	17	17	19	21	20	19	18
Total	15	15	21	28	34	46	48	51	52	55	62	63	67	68	73	81	84	88	95	104	104	104	114	131	140	145	148	152

Commodity PF (CPI): Tracks the weekly PF against the overall % of complete for each commodity.

Useful way to track the trends, which can lead to interventions

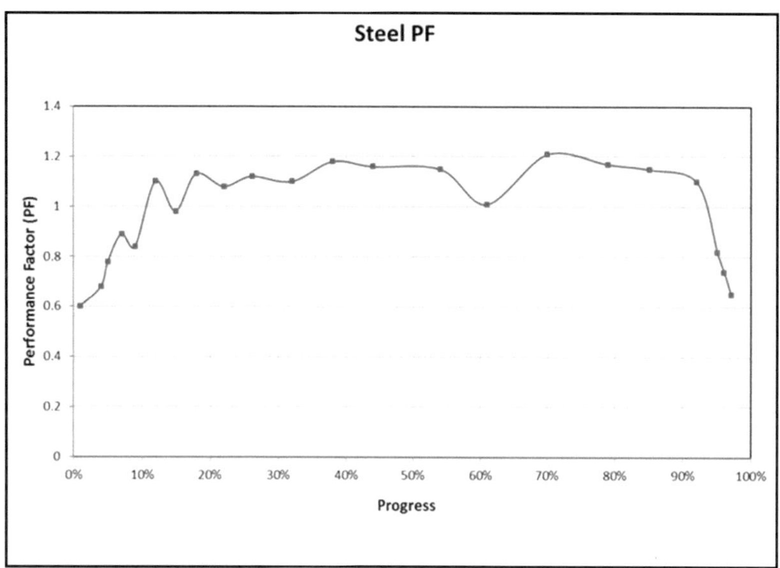

Commodity Installation Rates: Tracks the workhours per unit of measure for each commodity.

This process of measurement benchmarks the project performance against industry standards.

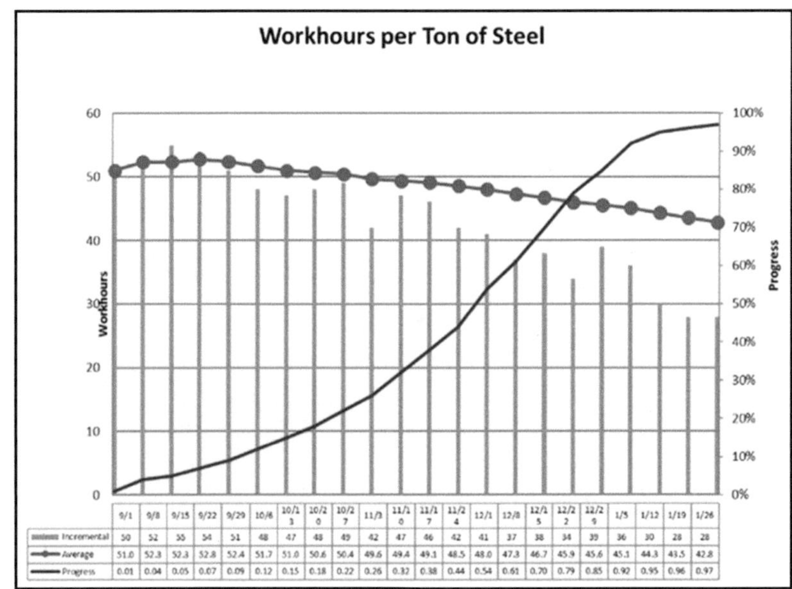

Weld Inches per Welder: Tracks the average number of diameter weld inches for a week against the average number of welders assigned to pipe and the total weld inches for the week.

Can also show the % complete overall and the total required.

At the end of the day, this is the number that counts. It is the pulse of your project and reporting it helps to drive productivity in all other sectors.

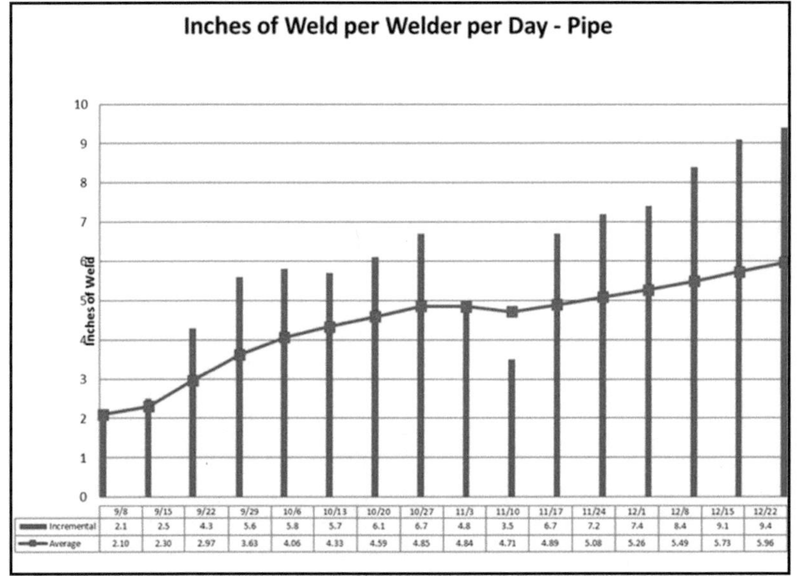

While the idea of tailoring the report function to be fit for purpose is a good thing, there are some cornerstone KPIs that I believe you must have if you are really going to understand the project:

- The IWP backlog,
- Weld inches per welder.

The IWP backlog tells you if your rate of planning is aligned with your burn rate (execution of work) and the Welder Inches tell you how good the project is at putting welds in front of welders, which is the heart of productivity.

Optional reports that we have also used in the past include, absenteeism and trade turnover, training, module completions and material shipments. You can expect that each of your projects will have a unique impediment to the flow of work that will need to be tracked and reported as an extension to this information.

Dashboard: Ideally the reports culminate into a dashboard, like the dashboard in your car, where fields that are green are compliant with expectations, fields in yellow indicate that there is slight variation from what would be considered normal and fields that are red need project management attention. This creates a stream of information that reads like a newspaper: Headlines that lead to high level summaries and then into the details if you want to read to the last paragraph.

Real Time Reports:

As I mentioned the hard part is setting up processes that produce the reliable data required to populate these reports. With a few exceptions, most of this data is produced as a by-product of Workface Planning managed in a WFP software environment. Data developed outside of the WFP software includes:

Weld inches per welder: Typically tracked by the Contractor's Quality Control department or the Welding Foreman.

Indirects as a % of Directs: Tracked by timekeepers through cost coded timesheets.

Scaffold performance: Backlog, earn rate and average cycle time is tracked in a scaffold database that is maintained by the scaffold team. Scaffold as a % of directs is tracked by the timekeepers through cost coded timesheets.

Delays: Extracted by the timekeepers from daily timesheets and tracked in a master spreadsheet.

RFIs: Tracked by the RFI Coordinator in a spreadsheet that is the master RFI log.

Data mining: To create an environment where this data could be mined, we developed a type of Data Warehouse called the **PRG Data-Lake** built in SQL. In my limited understanding of data management, I understand it to be a one stop shop where each of the departments submits their data in a standard format. Hosted on the project cloud it also pulls live data from the WFP software. A built-in report function allows the users to select filters of their choice: Window of time, Discipline, Area, Contractor etc. and then apply them to any of these reports.

The result is that for users who have been granted permissions, there is a live report on their phone based upon their own filters.

This truly is a paradigm shift, live data that shows the current reality of the project, from anywhere in the world that has phone service.

But it comes with a caveat: We have to learn how to handle the truth.

If we knee jerk to every little twist in the project then the people who generate the data will find some way to dilute the message and we will be back where we started.

The War Room: The Next challenge after you have determined what your weekly reports will look like is to consider how you will review them, which brings us to the concept of the 'War Room' (Command Center) that we have used quite successfully as the host for the weekly report.

It was a concept that we had used on projects with varying degrees of success, that was fine-tuned when we got the chance to work with some ex-army construction managers who showed us their refined model for tanks, troops and supply lines. We found that the logic of construction is very similar to military models with planning, resources, constraints and objectives all being very important in both fields. This is also where we learned that no plan survives first contact with the enemy, so you will need an underlying strategy to go with the plan.

The model for the war room that we had most success with had a plot plan of the project printed onto a magnetic whiteboard that was the table top in the board room. (perhaps sometime in the future this will become a hologram of the project). Then there were whiteboards, corkboards or computer monitors placed around the room for each of the disciplines: Safety, Quality, HR, Engineering, Procurement, Construction, Workface Planning, Project Controls and Turnover. With the Dashboard displayed at the front of the room as a summary. The departments were all responsible to maintain their boards and then to speak to them during the various meetings. This creates an information hub where project people could see the current state of the project at any given time.

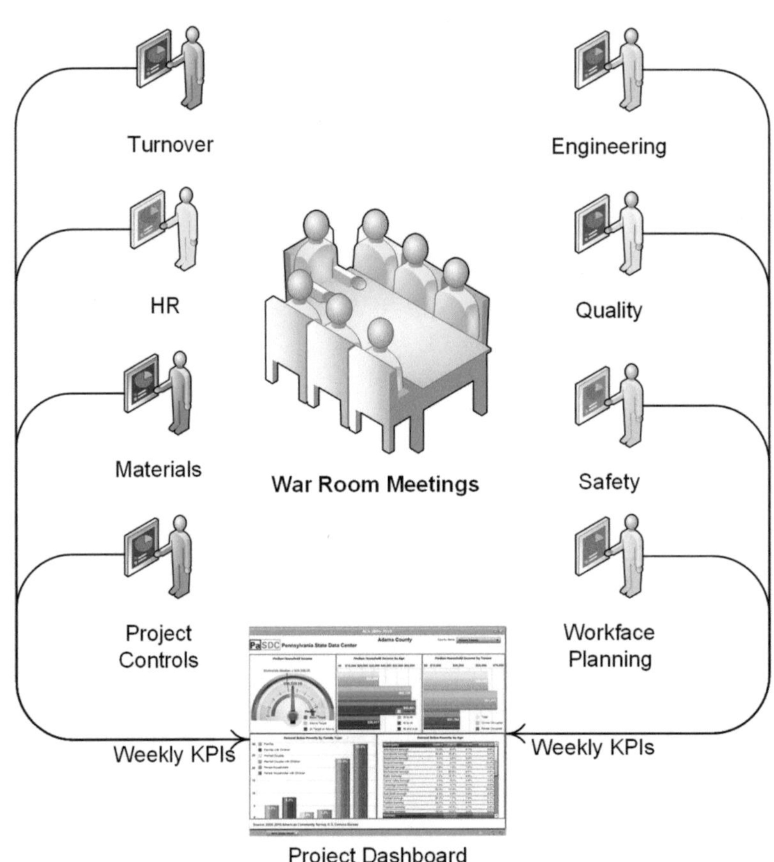

In Summary:

There are lots of studies and graphs that show how construction productivity is in decline or at least, not increasing at the rate of other industries. Knowing the complexity and dynamic nature of construction projects it is easy to understand how we struggle to continuously improve. The key to any process improvement model is to start with a stable environment and then to make incremental improvements that reduce the effort or increase the output. Industrial construction is anything but stable with unique designs and daily changing conditions executed by a constantly rotating workforce. This creates a very fluid environment that is difficult to manage day to day, even without the thought of improvement or optimization, or so it would appear.

The truth is that the way that we manage projects creates these dynamic conditions. We produce engineering in a general direction that roughly follows an undefined path of construction and do so without delivering complete engineering packages, due to a variety of reasons. This condition cascades and escalates through procurement and fabrication who optimize their own systems without consideration for construction so that the site ends up with only 7 or 8 parts out of every 10 that they need. Over many years this has created construction supervisors who had to learn to be agile. Known colloquially as cowboys, these industry professionals have become very good at problem solving on the fly, which creates its own problem. Their efforts buffer the impact of misaligned engineering, out of sequence procurement, a shortage of cranes, late scaffolds and permits that get issued an hour after we wanted to start work. In short, they are making progress despite our best efforts to derail the process, so it is very difficult to see the true impact of poor supply lines.

Just imagine what would happen if we were on their side.

Our current model for project execution is poised for a dramatic change. The same sort of change that we see around us when a condition becomes so intolerable that the status quo is just not good enough anymore.

Take the taxi industry as an example. Our society had become accustom to the overpriced monopolies that gave us no choice but to pay inflated fares. Then overnight Uber turned the industry on its head with a very simple model. Dramatic change caused by an acute need for a paradigm shift.

Imagine what will happen when we refine the AWP model to the point where we can predictably execute multibillion-dollar projects to land within 5% of budget and schedule because we know the productivity rates that will be achieved. Then, like any system that is stable, it would lead to continuous refinement of the processes and methods.

The truth is that we could be here now, if we consistently applied AWP and standardized the way that we measure productivity so that we could compare apples to apples and build statistics that accurately predict outcomes based on inputs.

Chapter 9: AWP Testimonies

The following AWP testimonies, from a collection of the industry's AWP SMEs are offered both as a demonstration of the wide spread application of the process and as an insight to some of the pot holes that you may encounter. The resounding message from these pioneers is that it is worth the effort and that it is not easy.

"King Solomon is quoted in the book of Ecclesiastes of saying *"that there is nothing new under the sun"* So, it is with WFP, my first day as a foreman was gut wrenching until an evening call to my Dad where he instructed me that I only needed to do 3 things. 1 – ensure my men have the right information the first time. 2 - do a MTO from the drawings and get their material as close to the work front as possible. 3 – ensure that the tools and equipment they require to install the material is at their work front. Then he said step back and let those boys do what they have been trained to do and they'll meet your schedule and beat your budget every time. That, was 26 years ago.

I have learned to look at this best practice through the lenses of PRINCIPLES and PRACTICES where principles are timeless and practices are timely. So, on every project no matter its location, industry or size from a principle perspective we are always doing WFP. However, the practice of how we execute the principle changes drastically in size, location and industry, and allows for a continuous improvement.

Understanding that consistency builds discipline, and that disciplined actions done consistently create success, then the biggest benefit I see today is a standardization of how we plan, delegate and track project completion.

We have come along ways since the early 2000's, but we still have a way to go."

Ben Swan
Element Construction
Nth America

"The key benefit in my world is the implementation of 3D modelling software. The development of a construction plan backed with a visual representation is an absolute must in my mind. This is because the project as a whole is not always able to envision what or how a plan on paper is able to be executed. We saw this on a recent project where a weekly meeting consisted of yelling and arguments, which accomplished nothing but headaches, until we were able to move the plans from the minds of Supervisors into a 3D visual that could be shared with everybody. This changed the mentality and attitude of the meetings into a team focus on one goal, executing the plan.

As a pipefitter by trade who transitioned into WFP/IM I was able to understand the importance of a detailed plan with strong focus on constraint removal to ensure there is limited delays in the field. This really came to light in my career when the industry slowed down a couple years ago and I had to go back to a pipefitting job for a few months to keep me busy. The company I went to work with had never heard of WFP or had any planning knowledge or interest in planning. I saw at this point that AWP/WFP is a key tool to assist in a projects success just as much as a strong safety and quality plan are."

BRETT HUNTLEY
AWP/IM Specialist
Nth America and Europe

"I had the opportunity to practice the job of Workface Planner Leader implementing the new principles of (WFP – AWP) - I will consider them as an entity since I consider Advanced Work Packaging the "extension" of WFP principles to the overall stages of a Construction Project – which are difficult to apply in the middle of the project. Key benefit of this process was the total study and identification of scope in each area, which helped tremendously in keeping available work fronts for our teams in the field. One of the biggest difficulties we dealt with was the denial of supervisors, superintendents and the construction teams to follow the philosophy and wait for the constraints to be removed. They are constantly seeking any kind of progress so they are willing to start in an area without all the requirements that assure efficiency can be met. Waiting for materials, scaffolds and other subcontractors is considered to be part of the normal construction process. Big breakthrough for us was that the teams that used the packages even without following 100% of the principles, found huge benefits and recognized that in the end it made a big difference. Plus, the area that used the WFP packages and waited for constraints to be removed was the most organized and achieved the highest percentage of completion in less time than all the other areas.

Before WFP there was no single person that took an overall view of the work to be executed from the view of all departments (material, quality, construction, etc.). The WFP packages made the building of the '4 week look ahead' accurate and sustainable. Before WFP it was just an unrealistic template with no connection to the work in the field.

The transition was very frustrating and challenging and since it was in the middle of the project a lot of targets were unrealizable. Nevertheless, it is common knowledge now that AWP has huge benefits. Next time we would definitely like to start earlier, choose Workface Planners with the proper mindset and skills and apply key components that will facilitate WFP procedures implementation."

CHRISTINA TSEPELI
Workface Planning Lead
Oil & Gas Construction Contractor
Europe

"I first worked with Geoff in 1999, when the study of industrial productivity was just getting started in Canada. Later we both worked on the inaugural COAA Workface Planning committee for almost three years during the research and development of the WFP model. I then used that model to help me set up WFP on a series of mega projects for a major EPC company.

We have come such a long way in WFP in the past 12 years but we have so much further to go. In the beginning the basic premise was to have IWPs no larger than 500 hours but contractors continue to build these sometimes in excess of 3,000 hours. These are much too large to manage.

While working in England this past year we were losing schedule and we had poor productivity. The client gave us permission to use Navisworks 3D software so that we could visualize the IWPs and show the field supervisors the advantages of utilizing WFP. When they arrived, the contractor was only obtaining 25% of their three week look ahead goals and within one month we changed this to 75% schedule compliance.

The AWP processes described in the first sections of this book work well if you have an EPC executing construction, but all too often this is not the case. Contractors are brought onto the project much too late and they struggle to get a backlog of IWPs, which leaves them poorly prepared to meet the fast schedule of today's projects.

Much can be done to improve WFP but it must be driven by the client. They are the ones that stand to gain the most from its implementation."

DAN GARON
Construction Manager
Major projects
Nth America and Europe

"The workface planning process is an essential piece of any successful mega construction venture within the projects environment, which increasingly dictates safe, quality results while accelerating schedules and continuing a capital efficient cost strategy. Advanced Work Packaging as a foundation for WFP is nothing new in my opinion. It does fill an obvious gap in the expertise level of the current project professions, that has formed in the construction industry due to the cyclic nature of the economics and the attrition rate. EPC contractors have, over time, lost their ability to retain the experience and knowledge of managing construction. Essentially the most capable EPC's on the books are in reality EP's with C being an after thought which is subcontracted at every level of the construction team. The AWP process provides a framework by which to build an efficient construction execution methodology that can bridge this gap, but needs implementing and to be fully supported in a timely and organized way. To reap the full benefits a project needs to implement AWP in late FEED early detailed design to ensure engineering is organized and focus to support predictable construction performance. Obtain early buy in from all key stakeholders and integrate into the contracts strategy to ensure accountability."

DANIEL LAUD
Project Controls Manager
Nth America and Europe

"I have been working in this area of specialty for over a dozen years and improvements are advancing in field productivity of the Industrial Construction Sector through the methodology and application of Advance Work Packaging. But in all honesty, it is still in the very early stages of acceptance with nothing but blue skies of opportunity ahead for those with vision. Until we reach the acceptance in industry that field productivity is as equally important as our current cultures in Safety & Quality, which have taken many years to develop. Today they are accepted has unquestionable departments of value added business units.

I do believe that we will proceed slowly but positively forward on AWP. Strong leadership on the Owner side along with partnership of Contractors coupled with educational tools like this book "Even More Schedule For Sale" are paramount. I would like to thank Mr. Ryan in his efforts to give back and move forward the industry with strong advance planning efforts, with the goal of seeing all interested parties benefit in a win/win partnership of successful field construction executions."

DENNIS MEADS
Industrial Engineer
Nth America and Europe

Being part of the team that pioneered the development and implementation of ConstructSim, which automates the best practice of Advanced Work Packaging, has given us at Bentley many valuable lessons learned. We now know that success in Advanced Work Packaging is 10% Technology and 90% Sociology. The technology portion of this equation is a critical enabler of data centric execution. The creation of Virtual Construction Models, leads to an integration of the project IT systems, which creates an environment that facilitates work pack creation, status visualization, look ahead planning, constraint analysis and change management.

We have also learned that it takes a programmatic approach combined with disciplined, rigorous project management leadership to make it successful. There is no big easy button, however our track record shows that an upfront investment of 1-2% of the total installed cost can yield a 10% reduction in overall cost and schedule, via productivity improvements and predictability.

As proof that AWP does work, we only have to look as far as the rich content of the submissions we receive in the Construction Innovation category of Bentley's annual 'Be Inspired' awards, as evidence that the effective implementation of AWP enables capital projects to be delivered on time and on budget!

AWP, as we know it today, has many forefathers: individuals, companies and industry groups like: COAA, CII and FIATECH that have worked tirelessly to bring the industry together. Building a collaborative pool of knowledge over the years that has led to AWP being identified as a formal best practice. I would encourage anybody who is serious about applying AWP to get involved with at least one of these organizations. My time as the Chair of the Construction Industry Institute's AWP Community of Practice has been rewarding and I feel that we are on the cusp of a digital revolution. What an exciting time to be in the construction industry!!"

Eric Crivella
Bentley Systems
Global

"I have been working on Workface Planning or Advanced Work Packaging projects for over 10 years now, starting in Canada and now in the USA. I have seen the good and the bad and I know that we have made some real advances in the organisation of engineering through our early involvement there. Over the ten years one of our biggest improvements has been to radically change the way that we handle documents, now we have one document control system, based on the cloud, which means only one version of each document.

Our next challenge is to bring fabrication up to the level of organisation that we need by making sure that they follow the right sequence and use the correct naming conventions."

JEFF FURLOTTE
AWP Specialist
Nth America

"We had the opportunity to implement WFP in a complex revamping project in Europe. The process had something to do with a cultural change that, even if imminent, it's far to achieve.

Even so, the benefits are many and easy to be cleared out.

Which are the benefits coming from more and better planning in construction?

The information is more accessible, and easier to manage and to communicate. The scope of work is clearer and easy to plan. The people within the organization speak a new common language and are aligned on what they need to discuss. And more, and more...

A final remark, WFP is about people and not about paper. It requires courage. It's a reasonable planning methodology made to serve the people that are working in the construction field making life easier to the people that are working in the office."

GREGORIO LABBOZZETTA
Project Field Engineer
Europe

"I had the pleasure of first meeting Geoff Ryan on a study visit to Alberta back in 2010. On this trip, I undertook a weeklong course in WFP under Lloyd Rankin and then attended the COAA conference where I met up with Geoff and undertook a few site visits. I could take away numerous learnings about WFP and assist companies here in Australia embrace a new way to deliver work in construction projects. Having applied WFP thinking into coal seam gas well construction and across numerous typical construction projects, the results speak for themselves. Along with productivity gains, the waste reduction and improved quality and schedule improvements are undeniable. Whilst the industry in Australia seems foreign when it comes to driving broad improvements across construction projects, there are a few real leaders that acknowledge the actual challenges in construction projects, and by embracing a proven planning technique such as WFP, change can happen for the better. The greatest benefit we are now seeing is that by having a fully unconstrained work pack, the coordinated planning process is a key focus as the enabler of constraint removal, and waste reduction is now front of mind for everyone."

LIAM STITT
DipEng; MMgt
MAIPM, Fellow SCLAA, UQ Industry Fellow, LCIAQ Council Member
Managing Director Essco-pl
Queensland, Australia

"When I joined the COAA committee in 2003, as the principle researcher, I could never have imagined that 15 years later we would see AWP being used in all corners of the world. When executed properly, I have seen AWP reduce capital projects Total Installed cost by at least 10%, reduce safety incidents to zero and improve schedule performance. The application of AWP is complex and difficult. It takes perseverance, faith and determination, but it works and it will be the way that projects are executed in the future.

Like any change the hardest part is getting started. Our experience hosting the AWP conference over the last 9 years has exposed us to Owners who are hungry for change being pursued by

Contractors who are eager to please. This makes the annual AWP conferences (awpconference.com) the ideal garden for many mutual-gain relationships."

Lloyd Rankin, MBA, PMP
Global AWP Specialist
ASI Group
Global

"I have been on numerous projects without AWP and it is the norm for them to go over budget and behind schedule. It has almost become an acceptable practice. So, how can we put an end to this and bring our projects back in the black and finish on time? I have applied Advanced Work Packaging (AWP) on several projects now and believe that it could be the answer. The earlier AWP can be implemented on our projects, the bigger the chance of success. We have proven time and again that AWP practices can save millions of dollars and keep the schedule from slipping too far to the right. There are several great 3D programs out there that can help manage and organize your documents and schedules, but without the expert manpower working behind the scenes, your IWPs will not be worth the paper they are printed on. Good quality Workface Planners are the driver for a company's AWP success."

LORNE SOOLEY
AWP Specialist
Nth America

"Advanced Work Packaging is, like Workface Planning, easy to say. Moreover, once folks look at what it is and its potential results, it makes sense. However, in our environment of instant gratification, it is hard to commit to, and an immense, thankless struggle to actually do.

I feel that the rest of the world is living in the Information Age and its high time that we recognize that we have the same potential in construction. The reality is that information management is an integral part of Advanced Work Packaging and it is our most challenging frontier.

Consider the following:

The answers to all the questions we ask on a project, depend on information; Where are materials, what is the scope, engineering status and construction progress. Somebody on the project already has this information and if we can find a way to pool this knowledge and then distribute it, everybody will know, which allows them to make informed decisions.

I know firsthand that we already have the capacity to create a cloud based single version of the truth on projects, so we don't need to know if it can be done, we just need people who have the will to do it. It will take determination, bravery, trust and most importantly, leadership to create an environment where data management is King."

Marco de Hoogh
Information Manager
Nth America and Europe

"I have spent the last ten years working with Geoff Ryan, first with Work Face Planning (WFP) and now with Advanced Work Packaging (AWP). I'm grateful that he has taken the time to put such a complex idea in simple terms on paper to push the change. This guide to AWP is explained carefully and thoroughly and I highly recommend everyone read it, regardless of where their organization is on the path of adoption of AWP processes.

We, as Hexagon PPM, are excited to help our customers realize the benefits of AWP. Although it originated in North America, we've seen interest rapidly growing internationally, particularly in the oil & gas, mining, and power industries. For our customers who have adopted this process, we've seen them improve time on tools by ten percent. It's incredible. Overall, the amount a company can save depends on how data-centric your execution is."

"It just makes sense"

Michael Buss
Senior Vice President
Hexagon PPM
Global

"It's been some time since Geoff Ryan and I worked at the same oil sands site and he gave me a copy of his first book. To say that it's been an eye opener complete with revelations of how to effectively deliver on schedule with quality, both essential to staying on budget is an understatement! When reading it, I experienced an epiphany which has been the guidance for all of my professional activities since.

The book describes in intimate detail the sequence of events and steps necessary to ensure that a project remains on track through the simple expedient of removing constraints which if not addressed and removed, would be obstacles in the path of progress.

What has been most instructive is the vast area within which this simple step (constraint removal) increases overall productivity, while reducing wait times for all concerned.

Implementing the basics of workface planning in industries other than construction has, in my experience, delivered the same outcomes, more effective detailed planning with better engagement of involved personnel, less waste time and resources, increased accuracy of materials received, issued and their ongoing status.

I encourage anyone working anywhere, regardless of which industry to get your hands on a copy; mine is well and truly a shabby copy because of its continual use. It's the only way that I know where you can strive for and achieve excellence in managing Cost, Quality and Schedule."

PAUL KALLAGHAN
Warehouse Coordinator
Northwest Redwater Partnership – Sturgeon Refinery
Alberta, Canada

"Organizations are discovering that critical review and revamping of processes and practices, along with aggressive training and mandates are not sufficient to realizing sustainable AWP. To me, the underpinning to realizing sustainable AWP lies in recognition, understanding and passionate drive to overcome resistance to change. The term ***Organizational Transformation,*** has been cited a AWP cornerstone, but there is a more fundamental mindset in the workforce that dictates the likelihood in realizing sustainable improvement. ***Change Readiness*** focuses on individual and group perspectives and biases throughout the enterprise, that can cause reversion to past practices and compelling AWP benefits to slip away. With this in mind, organizations are able to better understand the hidden landscape of significant obstacles to more effectively target efforts needed to embrace AWP over the long-term."

REG HUNTER
R.W. Hunter & Associates, LLC
Fiatech Program Dir. (retired)
URS/EG&G/LSI Productivity Improvement Systems Solutions Dir.
USA

"Advanced Work Packaging, in the words of someone who tried to put it down, is 'nothing new.' This should be comforting to those seeking to implement it. AWP does have a few new ideas, but work packaging is an established concept. AWP just updates it for the complexity of modern capital projects. There is no doubt that AWP, properly implemented, yields significant productivity gains and increased cost and schedule predictability. As with most things, the devil is in the details. AWP is a systemic approach to early planning through field execution. It takes consistent focus from many to succeed. Poor implementation of AWP will not yield benefits – those who have tried and failed are more likely to blame AWP than themselves. For those with some success, AWP is still a maturing practice and each organization is learning how to implement it within their culture and practices.

This book is not the last word as the industry improves, but it certainly has lots of good words. Thoughtful practitioners would be remiss not to study it."

WILLIAM J. O'BRIEN, PE, PhD
Professor
The University of Texas at Austin
Nth America and Europe

Chapter 10: The Future:

What does it look like?

My own opinion of the future is that the industry is poised for a massive revolution of Disruptive Innovation, along a fault line that is anchored on one side by traditional, stagnant, unpredictable construction, with an enormous buildup of pressure on the other side coming from a desperate need for innovation, accountability and reform. Every other industry has found ways to innovate and dramatically improve their productivity and yet industrial construction projects are stubbornly holding on to their traditional execution models, that we know don't work. How much longer do you think it will be acceptable for projects to run over cost and schedule because the project team did not want to apply 'known best practices'? How much longer will it be OK for project managers to plead ignorance or tradition?

My guess is that the game changing earthquake will be triggered by a ruling body, a government that dictates AWP for all public projects, or maybe it will be the financial sector that will no longer fund projects that are not regulated by AWP principles. Or maybe just a major oil and gas, power, mining or chemical company that implements mandatory AWP stage gates for all projects. The need is obvious, the AWP path is mapped and proven, as an industry we just need to mature to the point where we understand that poor project performance is a choice.

The model for disruptive innovation identifies a model for change based upon agile players filling the gaps left by the juggernauts who are too busy trying to make their old models work. This is probably what we will see in the industry where the old guard are so entrenched in their existing processes that they will not be able to adapt to the new game. This has certainly been our personal experience up to now, where our biggest successes have come from the smaller hungry new players.

Between stimulus and response there is a space. In that space is our power to choose and to grow.

Viktor Frankl was talking about our ability to be happy or at least to be in control of our own emotions, I liken this statement to being the adult in the room. When a teenager is having a hissy fit in your kitchen it is easy to engage in responsive behavior, pushing back as each point is made, leading to the ultimate losing statement of "my house and my rules". As the adult in the project management room it is your role to not respond like you feel, but to take the higher ground and accept that change might be a good thing, using logic rather than emotion. It is at this point that the right of true leadership will present itself and be yours to accept.

Our world is changing and you have the choice to step up or to be left to wonder in the desert. That doesn't mean that you have to like change, but you do have to get your head around it and understand that fighting change is a losing battle. You will find that it's exhausting trying to come up with reasons to stand pat, so my advice to you is to get on the train or get off the track.

So, what will the change look like?

Disruptive change, innovation gone wild, serial self-actualization, mutiny of common sense, change agents, paradigm uprisings, reorganization of the world order, failing for success, untethered access, positive destabilization, visioneering, breaking points that are tipping points, determination on steroids, redistribution of accountability, data fueled revolutions.

Facilitating the implementation of

Augmented Reality, Artificial Intelligence, Virtual Reality, Augmented Intelligence (Think of Jarvis from Iron Man) Paperless data centric projects, BIM, Lean, Agile, Pocket Super Computers, 3D printers, Construction Apps and Leap Motion.

All of these innovations are available now in some form and their intrusion into our world of project execution is only a matter of time, regulated by our access to data and our tolerance for change.

I understand that up to 25% of the jobs in the market now are filled by people who could not study for their careers, because the jobs they work in didn't exist when they were at college. Factor this complication by the average number of careers (four) that each of us have now multiplied by the rate of acceleration that we experienced between our grandparent's generation (typically one career), carried over to our grandchildren (10+ careers?) and you end up with the logic that most of the jobs in the future will be in fields that nobody has even thought of. And this is all happening in the next 20 years of our life time, which means that it is both an exciting and scary time to be alive.

The only constant is change.

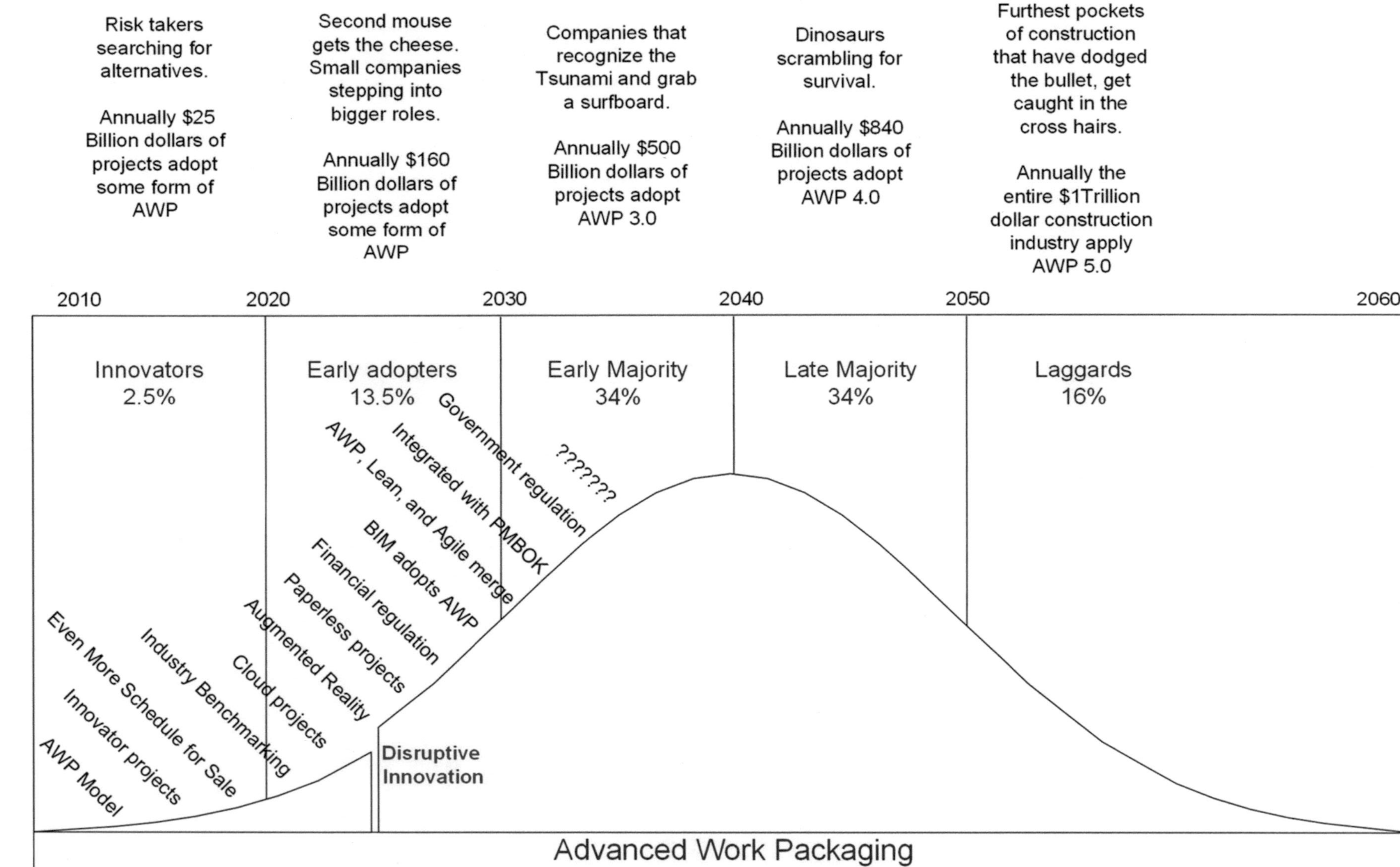
The Diffusion of Innovation Applied to AWP
Risk takers searching for alternatives.
Annually $25 Billion dollars of projects adopt some form of AWP
Second mouse gets the cheese. Small companies stepping into bigger roles.
Annually $160 Billion dollars of projects adopt some form of AWP
Companies that recognize the Tsunami and grab a surfboard.
Annually $500 Billion dollars of projects adopt AWP 3.0
Dinosaurs scrambling for survival.
Annually $840 Billion dollars of projects adopt AWP 4.0
Furthest pockets of construction that have dodged the bullet, get caught in the cross hairs.
Annually the entire $1Trillion dollar construction industry apply AWP 5.0
2010
2020
2030
2040
2050
2060
Innovators 2.5%
Early adopters 13.5%
Early Majority 34%
Late Majority 34%
Laggards 16%
AWP Model
Innovator projects
Even More Schedule for Sale
Industry Benchmarking
Cloud projects
Augmented Reality
Paperless projects
Financial regulation
BIM adopts AWP
AWP, Lean, and Agile merge
Integrated with PMBOK
Government regulation
???????
Disruptive Innovation
Advanced Work Packaging

The diffusion of innovation

Based upon our 1 trillion dollar (1000 $Billion) annual construction industry, I see the integration of AWP maturing through these stages.

2010 to 2020 - **Innovators:** $25 Billion per year in projects that adopt some form of AWP. These are the risk takers actively searching for alternatives. The industry sees some remarkable successes and starts to understand that the path to successful change is a combination of process, sociology and technology.

2020 – 2030 **Early adopters:** 160 Billion per year in projects that adopt some form of AWP. These are the bears, not the bulls, proceeding with the logic that the second mouse gets the cheese. Typically, small companies stepping into bigger roles or big companies that have learned to reinvent themselves. This is also the golden era of project management, sparked by a Disruptive Innovation event that turns the industry on it's ear and delivers on the promise of dramatic results. Technology becomes a standard feature for constructors, paper disappears and the cloud is the central platform for all data and communication. By the end of this period the AWP model will be on version 3.0 which will integrate the principles of Lean, Agile and BIM with the logic of work packaging in the framework of the PMBOK.

2030 – 2040 **Early majority:** 500 Billion per year in projects that embrace an advanced variation of AWP, which are mostly natural gas Cogen projects to generate the power that we need now that we are using electric cars, trains and aircraft. The companies joining the gold rush recognize that the tide has changed and agree to surf the Tsunami rather than drown. They are companies that work government projects or maintenance contracts, which are now regulated by rules of AWP engagement.

2040 – 2050 **Late Majority:** 840 Billion per year in projects now embrace some form of AWP. The Late Majority are the 'tethered dinosaurs', companies that reluctantly do what they are told to please the owner. They routinely fail AWP audits and only get projects because they have a salesman in a shiny suit or they are the only company available. AWP is on version 4.0 by the end of this period, where contractors have to maintain the asset for it's entire life cycle.

2050 – 2060 **Laggards:** 100% of the $1 Trillion industry is now mandated to apply AWP 5.0. The methodology is a standard format that is taught in high schools as Advanced Project Management. The last companies to come on board are the Ma and Pa teams from dark corners of the earth that had somehow dodged the bullets so far.

Now imagine what will happen when our trillion-dollar industry improves productivity by 25% and reduces the spend by $100 Billion. At the very least that makes another $100 Billion available every year to build more schools, hospitals, roads, water treatment plants, rail systems and environmental projects. We also know from experience that innovation feeds innovation so the rate of cost and schedule reduction will continue to accelerate through this period and we will continue to do more with less. On top of this, the technology that's introduced will keep driving costs down, wages up and safety incidents through the floor.

What's not to love?

The evolution of AWP.

In the last 10 years, we have seen Workface Planning mature into Advanced Work Packaging. This was a natural evolution from Installation Work Packages for the construction teams to engineering and procurement work packages that support the flow of construction. In the next ten years, I'm confident that we will see the integration of Information Management on the cloud as the glue that holds it all together.

In business, there is a continuous improvement standard that suggests that companies need to make a leap every 7 years, which fits with the logic that the economy also runs in 7-year cycles (since the great depression) and that people change their friends, taste buds and allergies every 7 years.

Whatever the cycle is, it is enough for us to know that the world of business runs through cycles and the survivors surf the waves and wade through the mud. Knowing that things do work in cycles we can predict when things are slow that they will get busy again and that when things are booming the industry will eventually slow down. Making use of our frontal lobe again, this would lead us to plan for the coming event: When times are slow get ready for the coming boom, when things are busy get ready for the quite times.

If this cyclical rejuvenation is true then we should expect a revision of AWP every 7 years or so, which has already started to prove out: 2005 - WFP then 2012 – AWP. I expect that the next revision to AWP will be subtler and may be an integration of some existing logic. It may be the fusion of AWP with information management and WFP software or perhaps the idea that Tool Time surveys are part of the model and not an optional extra. It could be the transition to standard commodity installation rates, or the blend of AWP principles with other proven project management processes. Either way the foundation of work packages for E, P & C has now been laid, let the evolution begin.

The integration of AWP with other project management initiatives is an interesting idea that is starting to grow legs. We already have a couple of projects where we have joined forces with Lean practitioners and achieved amazing results. Our most successful projects also had elements of Agile, BIM, PMBOK, Kaizen, Six Sigma and a good helping of Last Planner that helped them achieve their goals.

Which means that the next iteration of AWP maybe a unified model, where one or more of these systems is used to facilitate the overall philosophy of work packaging through E, P, & C.

So, what else is out there changing the world:

Agile: A collaborative scrum methodology (like the rugby scrum) that sets a series of incremental milestones for the team to achieve an overall result. Daily 15-minute check in meetings and month-long sprints achieve set targets. Allows members the freedom to innovate individually, while also working in teams.

BIM: Building Information Modelling: A process of developing intelligent 3D models for the management of commercial construction, where the details of the components are the model attributes. Select an object in the 3D model and you get the full spec including price. Can also store status (received, installed etc.).

Kaizen: Continuous Improvement that is based on the model for change: Identify the problem, Analyze the process, develop solutions, implement change, monitor the results, tweak the solution and repeat.

Last Planner: Work execution planning process where the Foremen develop work packages for themselves based upon: Can, Should and Will. (This was the foundation that evolved into AWP).

Lean: Plan - Do – Check - Act, then repeat. Continuous improvement that targets optimal productivity through the reduction of wasted time and materials. A very solid logic born from Dr. W. Edward Deming and his 14 points of total quality management, perfected by Toyota and now practiced worldwide.

PMBOK: Project Management Body of Knowledge is a technical project guide that identifies the inputs processes and outputs from 9 knowledge areas across the 5 stages of projects. In 2016 there were 710,000 certified Project Management Professionals (PMPs) across the world.

Six Sigma: is statistical modelling of data for the purpose of definitive quality control for specific targets. The target is to improve specific tasks within a process to the point where they have an error rate that falls within a tight band of acceptance.

7 Habits of Highly Effective People: More of an ideology than a process, this is the backbone of business sociology that will be required to guide the change that is coming. Mandatory study for anybody who is serious about working in the field of Advanced Project Management.

For those practitioners amongst you, I don't mean to slight any of these processes by not giving a detailed account of their elements. I do want the readers to understand that there are existing models for improving project results and that they are not competing with AWP, they are complimentary. We are all trying to improve results and these processes all hold different pieces of that very complex puzzle. And, (boldly going into a snake pit of controversy): No single process has all the answers to our problem of optimal performance and predictable project results, and this includes AWP. The right answer is going to be a blend of these and other processes that will evolve based on empirical data, which will probably give us a new hybrid variation every 7 years or so.

Predictably the new model will be a blend of the strengths from these other systems built on the platform of package creation and execution, that is the backbone of AWP. Certainly, we already see elements of each of these systems in the AWP model. Our challenge is to develop a simple system that blends the best parts of these known 'best practices' into a working model, that other people can execute.

And herein lies one of the obstacles that we face, the logic that we must reinvent the wheel and that there is some system out there, yet undiscovered, that will revolutionize the world. I don't think that there is, I think that we already have the answers, hidden in these systems, but we don't fully understand the questions.

All of these systems were purpose built to fill process gaps for specific problems. But problems being as unique as they are, one size solutions don't fit all. So, the key to developing the answer is to fully understand the gap between the current reality and the desired state and then to shop the menu of systems or parts of systems, to find a combination that will address the problem.

The core problem for industrial project execution is a misalignment between Engineering, Procurement and Construction, which created a current reality of: construction teams scrambling to find successive work fronts. AWP has been specifically designed by the industry to address this gap by introducing the logic of work packages in a one to one correlation across E, P & C. However, AWP is only the foundation and the systems, work processes and cooperation that are needed to make this functional is the hard stuff. Those solutions will have to come from our tool box of project management techniques.

We are made wise not by the recollection of our past but by the responsibility for our future.

George Bernard Shaw.

Summary and Links:

Well I hope that you found something in the book that you can use and apply. If my calculations are right then you also found somethings that you disagreed with. I would enjoy nothing better than to be proven wrong, because that means that we will have a new right.

We only know what we are taught or have experienced, so I would encourage you to share your experience and also learn how to learn from others.

The perfect environment to do this is in one of the collaboration organizations:
Construction Owners Association of Alberta: https://www.coaa.ab.ca
Construction Industry Institute: https://www.construction-institute.org
Fiatech: http://fiatech.org
Curt: https://www.curt.org
ECC: http://www.ecc-conference.org
Lean Construction Institute Australasia: http://www.lcia.org.au
Advanced Work Packaging Institute: https://www.workpackaging.org
Linkedin AWP groups:

Support for AWP applications can also be provided by:

ASI Group (also conduct the annual conferences): https://www.groupasi.com
Bentley ConstructSim: https://www.bentley.com
Construct-X: http://www.construct-x.com
Element Industrial Solutions: http://elementindustrial.com
Hexagon (Intergraph) Smart Construction: http://hexagonppm.com
Insight-awp (us): www.insight-awp.com

Academia that are involved in the exploration and education of AWP.

University of Queensland
University of Alberta
University of Calgary
University of California, Berkley
University of Houston
University of Texas at Austin

Also check out the tutorial videos on youtube.com

Printed in the United States
by Baker & Taylor Publisher Services